GUS Protocols

GUS Protocols: Using the GUS Gene as a Reporter of Gene Expression

Edited By

Sean R. Gallagher
Hoefer Scientific Instruments
San Francisco, California

ACADEMIC PRESS, INC.
Harcourt Brace Jovanovich, Publishers
San Diego New York Boston London Sydney Tokyo Toronto

Front cover photograph: Cross section (15μ) of a loblolly pine *(Pinus taeda L.)* cotyledon showing transient gene expression from a microprojectile bombardment experiment. Photograph was taken using polarized light and interference contrast on a Leitz Diaplane microscope. Courtesy of Dr. Anne-Marie Stomp, Forestry Department, North Carolina State University, Raleigh, North Carolina.

This book is printed on acid-free paper. ♾

Academic Press, Inc.
San Diego, California 92101

United Kingdom Edition published by
Academic Press Limited
24–28 Oval Road, London NW1 7DX

Library of Congress Cataloging-in-Publication Data

GUS protocols : using the GUS gene as a reporter of gene expression /
edited by Sean R. Gallagher.
p. cm.
Includes bibliographical references and index.
ISBN 0-12-274010-6
1. Beta-glucuronidase genes. 2. Gene expression. 3. Biochemical markers. 4. Genetic engineering--Technique. I. Gallagher, Sean R.
QH447.8B46G87 1991
574.87'3224--dc20 91-28623
CIP

PRINTED IN THE UNITED STATES OF AMERICA

92 93 94 95 96 97 EB 9 8 7 6 5 4 3 2 1

Contents

Contributors

Numbers in parentheses indicate the pages on which the authors' contributions begin.

Roger N. Beachy (127), Department of Cell Biology, Scripps Research Foundation, La Jolla, California 92037

Ben Bowen (163), Department of Biotechnology Research, Pioneer Hi-Bred International Inc., Johnston, Iowa 50131

Thomas B. Brumback, Jr. (77), Department of Data Management, Pioneer Hi-Bred International Inc., Johnston, Iowa 50131

Stuart Craig (115), CSIRO Division of Plant Industry, Canberra, ACT 2601, Australia

Leigh B. Farrell (127), CSIRO Division of Plant Industry, Canberra, ACT 2601, Australia

John N. Feder (181), Howard Hughes Medical Institute, and Department of Physiology and Biochemistry, University of California, San Francisco, California 94143

E. Jean Finnegan (151), CSIRO Division of Plant Industry, Canberra 2601, Australia

Pamela Flynn (89), Department of Biotechnology Research, Pioneer Hi-Bred International Inc., Johnston, Iowa 50131

Wolf B. Frommer (23), Institut für Genbiologische Forschung, D-1000 Berlin 33, Federal Republic of Germany

Sean R. Gallagher (1, 47), Hoefer Scientific Instruments, San Francisco, California 94305

Daniel R. Gallie (181), Department of Biochemistry, University of California, Riverside, California 92521

Nancy Galvin (189), Department of Pathology, Saint Louis University School of Medicine, Saint Louis, Missouri 63104

Jeffrey H. Grubb (189), Department of Biochemistry and Molecular Biology, Saint Louis University School of Medicine, Saint Louis, Missouri 63104

Stephen G. Hughes[1] (7), Nuovo Crai, Caserta, Italy
Sabine Hummel (23), Institut für Genbiologische Forschung, D-1000 Berlin 33, Federal Republic of Germany
Richard A. Jefferson (7), CAMBIA Organizational Office, Lawickse Allee #22, Wageningen 6707 AG, The Netherlands
John W. Kyle (189), Department of Medicine, Section of Cardiology, University of Chicago, Chicago, Illinois 60637
Thomas Martin (23), Institut für Genbiologische Forschung, D-1000 Berlin 33, Federal Republic of Germany
John J. Naleway (61), Molecular Probes, Inc., 4849 Pitchford Avenue, Eugene, Orgon 97402
Jane K. Osbourn (135), Department of Virus Research, John Innes Institute, AFRC Plant Science Research Centre, Norwich NR4 7UH, United Kingdom
A. Gururaj Rao (89), Department of Biotechnology Research, Pioneer Hi-Bred International Inc., Johnston, Iowa 50131
Anne-Marie Stomp (103), Forestry Department, North Carolina State University, Raleigh, North Carolina 27695-8002
Carole Vogler (189), Department of Pathology, Saint Louis University School of Medicine, Saint Louis, Missouri 63104
Virginia Walbot (181), Department of Biological Sciences, Stanford University, Stanford, California 94305
Lothar Willmitzer (23), Institut für Genbiologische Forschung, D-1000 Berlin 33, Federal Republic of Germany
Kate J. Wilson[2] (7), Wye College, University of London, Wye, Ashford Kent TN25 5AH, United Kingdom
T. Michael A. Wilson (135), Center for Agricultural Molecular Biology, Cook College, Rutgers University, New Brunswick, New Jersey 08903
Rosa-Valentina Wöhner (23), Institut für Genbiologische Forschung, D-1000 Berlin 33, Federal Republic of Germany

[1] Present address: Plant Breeding International, Maris Lane Trumpington, Cambridge CB2 2LQ, United Kingdom

[2] Present address: Department of Microbiology, Wageningen Agricultural University, Hesselink van Suchtelenweg 46703 CT, Wageningen, The Netherlands.

Foreword

"Nothing great is achieved without chimeras."
Ernest Joseph Renang
L'Avenir de la Science [1890]

Gene fusions were used in biological research well before the advent of recombinant DNA. The pioneering work of Beckwith, Signer, and their colleagues in the middle to late 1960s (Beckwith *et al.*, 1967; Miller *et al.*, 1970) in juxtaposing the *trp* and *lac* operons of *Escherichia coli* started a revolution in the analysis of biological function. The concept of using one gene with a product that is easy to detect—in this case β-galactosidase—to infer the behavior of another gene that is functionally fused to it was so powerful in its simplicity and so pervasive in its applications that there has been little recognition of the fundamental paradigm shift that it entailed. For the next two and a half decades, as recombinant DNA technologies emerged and as transformation systems were developed, thousands of scientists used and continued to develop the LAC system. The information obtained with these gene fusions in many cases could not have been obtained by any other methodology. The LAC system has tended to find its greatest use in the major laboratory model systems, such as *E. coli, Saccharomyces cerevisiae, Drosophila melanogaster,* in mammalian cells in culture, and lately in transgenic mice. The principal restriction of its use has generally been due to either a lack of suitable transformation methods for a particular system, or the presence of an endogenous β-galactosidase activity. These restrictions have been particularly troublesome in plants and other agricultural systems. Notwithstanding these limitations, the LAC system has proven the extraordinary power of gene fusions in prokaryotes, fungi and animals, and has clearly demonstrated that the development of a single tool for experimental manipulation when broadly applied to numerous systems, can have a profound effect on the quality and type of science that emerges.

Agricultural research, especially plant molecular biology, had not been able to benefit greatly from the gene fusion paradigm until the advent of the first transformation methods for plants in 1983–1984 and of the GUS system in 1987–1988. The GUS system has been instrumental in facilitating the development of novel transformation methods for many crops, and in allowing detailed analysis of gene action in transgenic plants, bacteria, and fungi of agricultural importance. In particular, use of GUS has facilitated the routine experimental manipulation and analysis of gene action in two and three dimensions. Cell-, tissue-, and organ-specific gene expression is now routinely studied using GUS fusions in transgenic plants.

However, agricultural sciences have only just begun to exploit the important experimental paradigm of gene fusion approaches due to the intrinsic complexity of the systems studied. The gene fusion paradigm as it exists now—at least with regard to its more sophisticated applications—is still largely shaped for model systems where environmental variation and interactions are minimized. Much of plant biology, especially agricultural plant biology, cannot be readily reduced to such model systems. For agriculturalists, performance in highly complex ecosystems is the ultimate criterion for success, and extrapolation from simple models cannot be expected to deal with this enormous leap.

We are now in the midst of the development of at least two more paradigm shifts that may be specific to the agricultural research agenda and that may be required to bridge the gap between model systems and agriculture. One of these can be considered largely scientific in nature, the other apparently social. These shifts are related to what type of science can be done and who gets to do it.

The first of these paradigm shifts is necessitated by the emergence of molecular biology from the confines of laboratory walls into the real and variable conditions of farmers' fields and represents a challenge of substantial magnitude. The requirements that gene action and all its consequences be understood under uncontrolled conditions, where the vicissitudes of the environment must be accepted not as a confounding influence but as a natural and necessary component of an experimental system are daunting. And yet the imperatives of agricultural improvement and the onward rush of transformation methods for crop plants, where performance of the engineered crop in the field is critical, have hastened the application of new approaches to the field without yet achieving the change in thinking and experimental methodology that will be necessary to do it sustainably, intelligently, and equitably. We need methods and ways of approaching problem-solving that can be used in the complex environment that dictates the biology and ultimately the performance of crops that give us new insights into the interactions between organisms and between organisms and their environment. We must ask whether these new methods can be expected to emerge as incremental improvements or whether there must be completely new approaches.

The other paradigm shift is related to who is empowered to ask questions and use these new technologies, as this ultimately determines who benefits from them. The most severe agricultural problems tend to occur where scientific research is most difficult and where solutions need to be found locally, typically in less developed countries. These problems are often socially, environmentally, and economically devastating. It is imperative, therefore, to design and distribute tools and methods that can function well in these situations and that will empower sophisticated local scientific research to address and overcome local problems.

The GUS system represents just one small step towards the necessary paradigm shifts. But it is an exciting one because it is a step that seems to be capable of very significant extension and improvement as the reconciliation between the needs of agriculture and the environment and the tools and priorities of the laboratory scientist proceeds. The possibilities for continued extension and development of the GUS system in new directions lie in the biological processes in which the enzyme is involved.

The Biology of GUS Gives Hints for New Possibilities

All vertebrates detoxify and excrete the myriad superfluous compounds that their systems encounter by one of a small number of mechanisms. These compounds will include plant secondary compounds and animal metabolites, hormones, and other endogenous compounds that are excess to requirements, and other xenobiotics, including drugs. The prevalent mechanism in most vertebrates is to conjugate these compounds with a water soluble handle, most commonly glucuronic acid, and then to excrete these conjugates through the circulatory system and ultimately through urine and bile. *Escherichia coli* has evolved as a key component of the endosymbiotic intestinal flora of most vertebrates and has developed the ability to metabolize these numerous and diverse compounds through the use of GUS and the related functions encoded by the *gus* operon (see Chapter 1), and can use the released glucuronic acid as both a carbon and energy source.

Within plants and other organisms such as fungi and insects that lack circulatory systems, conjugation with glucuronides is not generally used. Instead, these organisms typically conjugate xenobiotics with glucosides, and sequester them rather than excrete them. Additionally, β-D-glucuronides are not typically found as important structural or biochemical components of plants, and thus GUS activity in these organisms or in the organisms that are associated with them is not expected nor generally observed.

An understanding of this pathway for metabolism of glucuronides and the appreciation of its ubiquity in vertebrates and relative absence in plants raises the novel and exciting possibility of manipulating the pathway and its components to design tools that can deal with several of the key restrictions and limitations of current gene fusion technology. Now that many tens of thousands of transgenic plants representative of dozens of genera have been generated that express GUS, we can approach this task with some confidence in its innocuous effects and ubiquitous utility.

Toward *in Vivo* and *in Campo* Molecular Tools

The first and most pressing limitation is that analysis of GUS is still performed *in vitro,* in destructive assays of extracts or through histochemical analyses of dead tissues. In the development of transformation methods as well as in the analysis of gene action, the ability to analyze GUS activity nondestructively, *in vivo,* would dramatically expand our capabilities. It would also provide a major step toward developing *in campo* molecular biology—stimulating analysis of gene action and its resulting biology under field conditions.

There are two clear routes to achieving *in vivo* GUS analysis, with substantial progress being made on both fronts. The fundamental problem with *in vivo* analysis is that GUS substrates—β-D-glucuronides—are highly water soluble and therefore cannot readily traverse the lipid bilayers of living cell membranes. Additionally, enzymatic analysis, if it is to be quantitative, must be carried out in significant excess of substrate. Therefore, the resolution can be considered as either of two complementary ways of getting enzyme and substrate together; by bringing the enzyme out of the cell so it can interact with substrates in the extracellular spaces, or by transporting the substrates across the plasma membrane where they can be acted on by cytoplasmically localized GUS.

The first route derives from the observations of Iturriaga, Kavanagh, Schmitz, and their colleagues that GUS could readily and efficiently traverse many of the membrane systems within plant cells when fused to the appropriate signal or transit peptide (Kavanagh *et al.*, 1988: Schmitz *et al.*, 1990; Itturiaga *et al.*, 1989). In particular, Gabriel Iturriaga's work showed that when GUS was fused to the signal peptide from the potato tuber protein, patatin, the resulting hybrid was targeted through the endoplasmic reticulum of transgenic plants, where the protein was stable, although greatly diminished in enzyme activity. He further showed that the diminution of enzyme activity was due to an *N*-linked glycosylation of GUS, which could be prevented, with concomitant restoration of enzyme activity, by addition of tunicamycin, an antibiotic known to block glycosylation. These observations were extended by Leigh Farrell and Roger Beachy (this volume, Chapter 9), who have altered the sequence that renders GUS susceptible to this glycosylation, without losing enzymatic activity. This step, and subsequent developments that will come soon as our understanding of protein targeting in plants matures, should allow efficient secretion of GUS and its retention within the extracellular space. It is then obvious that application of highly water-soluble GUS substrates should result in cleavage and detection (or bioactivity) in and on living plant tissue. Issues remaining to be resolved will include maintenance of enzyme localization near or at the site of enzyme synthesis (probably achieved by fusion with components of cell wall proteins or simply by virtue of GUS' large tetrameric structure), transport of the substrate to the site of enzyme localization (GUS substrates are most likely highly phloem translocatable, but probably not cuticle permeant), and migration of the product of the enzyme reaction from the site of cleavage (not a serious problem with current histochemical substrates, but an important criterion for the application of fusion genetics or novel fluorochromes). A number of these steps can be dealt with by sensible design of substrates, which is very straightforward.

The second route toward *in vivo* analysis is more ambitious, but will ultimately be more general, quantitative, and powerful and involves the manipulation of the membrane transporter for β-D-glucuronides. As described by Kate Wilson and her colleagues (Chapter 1), and in unpublished work by Weijun Liang, Peter Henderson, and their colleagues at Cambridge University (W.-J. Liang, P. Henderson, T. J. Roscoe, and R. A. Jefferson, unpublished), and building on the excellent thesis work of Francois Stoeber in the late 1950s (Stoeber, 1961) the glucuronide permease from *E. coli* is now being characterized molecularly and biochemically and seems to be a very exciting candidate for providing a general mechanism for actively transporting diverse GUS substrates into living cells. The permease is a single polypeptide proton symporter that can accumulate a very wide range of GUS substrates into *E. coli*. The manipulation of this gene to allow functional expression in transgenic eukaryotes is a likely development in the near future. When this is achieved, application of any of a number of fluorogenic, chromogenic, or bioactive substrates may then allow true, quantitative *in vivo* analysis and fusion genetics. One of the most attractive features about developing the permease for this purpose will be its ability to concentrate substrates from very low external concentrations to levels well in excess of the K_m value for GUS cleavage.

Toward Fusion Genetics

Both of these approaches to developing a viable assay system will also serve to hasten the development of effective fusion genetic methods. Fusion genetics is one of the most powerful components of the gene fusion paradigm as it was originally elaborated in bacterial genetics, but has been very poorly developed for complex eukaryotes, with some notable exceptions (e.g., Bonner *et al.*, 1984). The basic premise that underlies fusion genetics is that one can perform not only visualization or measurement of gene fusion activity, but genetic selections for variants in activity of the gene fusion, thus selecting for and obtaining both cis- and trans-acting mutations. These mutations can then be used to infer relationships between the controlling elements and components of the host genome. In bacteria, for example, this has been done in countless instances by fusing the *lac* operon to the controlling sequences of another gene (a classical operon fusion) and then selecting for constitutive or otherwise inappropriate expression of LAC. The mutations affecting the control of the fusion have then been used to define pathways of gene control.

This tool will be particularly important in complex plants where traditional genetic analysis is cumbersome relative to microbial genetics, due to long generation time, large size, genomic complexity, and developmental and spatial variations of cell types and gene action. Now that *Arabidopsis thaliana* is emerging as a powerful model system, fusion genetics, when developed in a versatile and general way, will become even more elegant and productive. Development of fusion genetic methods using the GUS system will involve the synthesis of novel GUS substrates which will yield bioactive products upon GUS action. This synthesis is almost the exact converse of what vertebrates do; instead of rendering bioactive compounds inactive by conjugating them to glucuronic acid—which is the vertebrate detoxification rationale—we are proposing to render these conjugates active by applying them to transgenic organisms that express GUS.

It is thus obvious that design and preparation of substrates can also make use of the vertebrate conjugation system, on which there is an extensive literature (Dutton, 1966, 1981). Several thousand glucuronide conjugates have been described from the urine or feces of animals. These are mostly biologically inactive derivatives of highly bioactive substances. Chemical synthesis or biological preparation of such compounds could be tailored to show particular effects on transgenic plants only upon cleavage of the conjugate by GUS. Almost any compound with an O-glycosidic linkage in the β configuration to glucuronic acid is a substrate for GUS. One can envision plant growth regulators or toxins, for example, that could manifest their biological activity on a transgenic plant only if that plant, or some subset of the cells of the plant, was expressing GUS. This could thus provide genetic selections either to plants or cells that acquired mutations that altered the GUS expression in a particular manner. It would also provide a route to modification of the systems' biology in a conditional manner. The possibilities are limited largely by the ingenuity of the investigator.

Summary

It is clear that there are numerous and exciting possibilities for future developments of GUS to provide novel tools. It is also clear that these tools, and tools like them, need to be developed with particular constraints in mind, including the constraints of nonmodel system biology, such as agriculture, and the constraints of the resource-limited scientist. Those who have been "GUS-conscious" for some time will recognize my exhortations regarding *in vivo* analysis and fusion genetics from previous discussions, talks, or papers, and will justifiably ask why, if these methods are so near and so powerful, do they not yet exist? The answer certainly lies in the extraordinary number of questions that are already accessible by use of the first generation of the GUS system as a simple analytical tool and the limited time and energy that can be expended for tool development when substantial answers to these questions can be obtained now. Our scientific sophistication is growing as we absorb the answers that early gene fusion experiments are providing and appreciate the limitations of both the questions and the tools. The next generation is almost upon us, and the questions will be fun indeed.

The GUS system is meant to be widely available. Those wishing clones or vectors for research should feel free to write to the author at CAMBIA to obtain these samples free of charge.

"Let's see what's out there."
Jean Luc Picard, 1990

Richard A. Jefferson

References

Beckwith, J. R., Signer, E. R., and Epstein, W. (1967). Transposition of the *lac* region of *E. coli*. Cold Spring Harbor Symp. Quant. Biol. **31**:393.

Bonner, J. J., Parks, C., Parker-Thornberg, J., Mortin, M. A., and Perlham, H. R. B. (1984). The use of promoter fusions in Drosophila genetics: isolation of mutations affecting the heat shock response. *Cell* **37**:979–991.

Dutton, G. J., ed. (1966). "Glucuronic Acid, Free and Combined." Academic Press, New York.

Dutton, G. J. (1980). "Glucuronidation of Drugs and Other Compounds." CRC Press, Boca Raton, Florida

Helmer, G., Casadaban, M., Bevan, M. W., Kayes, L. and Chilton, M.-D. (1984). A new chimaeric gene as a marker for plant transformation: the expression of β-galactosidase in sunflower and tobacco cells. *Bio/technology* **2**:520–527.

Iturriaga, G., Jefferson, R. A. and Bevan, M. W. (1989). Endoplasmic reticulum targeting and glycosylation of hybrid proteins in transgenic tobacco. *The Plant Cell* **1**:381–390.

Kavanagh, T. A., Jefferson, R. A. and Bevan, M. W. (1988). Targeting a foreign protein to chloroplasts using fusions to the transit peptide of a chlorophyll a/b protein. *Molec. Gen. Genet.* **215**:38–45.

Miller, J. H., Reznikoff, W. S., Silverstone, A. E., Ippen, K., Signer, E. R. and Beckwith, J. R. (1970). Fusions of the *lac* and *trp* regions of the *Escherichia coli* chromosome. *J. Bacteriol.* **104**:1273.

Schmitz, U. K., Lonsdale, D. M. and Jefferson, R. A. (1990). Glucuronidase gene fusion system in the yeast, *Saccharomyces cerevisiae*. *Curr. Genet.* **17**:261–264.

Stoeber, F. (1961). Etudes des proprietes et de la biosynthese de la glucuronidase et de la glucuronide-permease chez *Escherichia coli*. These de Docteur es Sciences, Paris.

Preface

GUS Protocols: Using the GUS Gene as a Reporter of Gene Expression introduces researchers to the diverse applications of the GUS gene fusion system. Originally developed by R. A. Jefferson and co-workers, the importance of β-glucuronidase, or GUS, as a reporter of gene expression in plants is illustrated by recent literature citations. In the first sixth months of 1991 alone, the assay was cited over 120 times. The GUS assay also produces visually striking results, as shown by both the cover and the color plates of this book. These beautiful color photographs illustrate the blue histochemical staining of GUS activity in transgenic plants, tracing patterns of expression of tissue-, organ-, and developmentally-specific genes.

The GUS system is popular because of its simplicity, versatility, and robustness. To use the GUS gene fusion system, typically the *Escherichia coli gusA* gene is coupled to a gene of interest. Once this fusion is introduced into plant (or animal) cells, GUS expression reports activity of the other gene. Because plants normally lack endogenous GUS, detection of the GUS enzyme either in whole tissue or tissue homogenates provides a sensitive background-free measure of gene expression. GUS from *E. coli* is also a good reporter enzyme in animal cells in spite of endogenous GUS activity.

GUS Protocols is divided into five sections covering the general use of the GUS reporter gene, details of the GUS assay, histochemical detection, applications of GUS in plants and, finally, animal genetic analysis. As a reporter of gene expression, the gene isolated from *E. coli* is almost exclusively used. To gain insight into the origins and other potential uses of the *gusA* (formerly *uidA*) gene from *E. coli,* a detailed discussion of the *gus* operon is presented in the first chapter of the book. This is followed by an introduction to the various assays required to effectively analyze GUS activity, including chapters on the properties of GUS substrates and laboratory exercises for biochemistry classes.

Frequently, researchers must assay huge numbers of samples while analyzing transgenic plants. Two chapters address this, one describing automated tissue grinding and extraction and the other GUS assays with microplate fluorescence readers. Several chapters deal with histochemical localization of GUS in plants and animals, and these feature extensive discussions of the problems often encountered. The number of applications that rely on the GUS system is immense, and this kaleidoscope of research is discussed in the last part of this book with chapters covering analysis of secretory systems, molecular plant virology, transposable elements, and genetic analysis of animal cells and tissue.

Variations on the GUS gene fusion system will continue to be developed, providing new and novel applications (see Foreword). However, because this book provides a foundation made of key references, techniques, and procedures, *GUS Protocols* will also continue to stay an invaluable resource well into the future.

Without the help from both organizations and individuals the publication of *GUS Protocols* would not have been possible. I am grateful for the support and assistance provided by the following companies: Pioneer Hi-Bred, Inc. and Clontech Laboratories for generous financial contributions to offset the cost of the color plates found in this book; Hoefer Scientific Instruments for making the inception and continued work on this book a possibility; and Academic Press for their persistence and determination. Special thanks goes to Lorraine Lica of Academic Press for her skillful guidance in preparing *GUS Protocols*.

Introduction

Sean R. Gallagher
Hoefer Scientific Instruments
San Francisco, California

β-Glucuronidase, or GUS, has in the last 5 years become the reporter enzyme of choice for plant genetic research. The gene for GUS currently used was originally isolated from *Escherichia coli* by R. A. Jefferson and co-workers (Jefferson *et al.*, 1986, 1987; Jefferson, 1987; Wilson *et al.*, Chapter 1, this volume) and is now commercially available in a variety of configurations that link the GUS gene to the promoter, gene, or signal sequence of interest (see appendix for plasmid constructs). A full description of the *E. coli* β-glucuronidase system and its analysis in *E. coli* and other bacteria is presented by Wilson *et al.* (Chapter 1). Through use of fusion plasmids in both plant (Martin *et al.*, Chapter 2) and animal systems (Gallie *et al.*, Chapter 13; Kyle *et al.*, Chapter 14), the activity of intact or altered promoters and the regulation of genes during development are easily studied. Other applications include recovering transformants without the use of antibiotic selection (Summerfelt *et al.*, 1991), cytotoxin research (Koning *et al.*, 1991), protein targeting (Farrell and Beachy, Chapter 9), pseudovirus expression and molecular plant virology (Osbourn and Wilson, Chapter 10), and assays for transposable elements (Finnegan, Chapter 11).

The *E. coli* enzyme has a monomeric molecular mass of approximately 68,000 and is encoded by the *gusA* (previously referred to as the *uidA*) locus (Wilson *et al.*, Chapter 1). The mammalian enzyme has a monomeric molecular mass of 75,000–82,000 and is, like GUS from *E. coli*, a tetramer. The gene for the mammalian enzyme has also been cloned from both human and mouse (Kyle *et al.*, Chapter 14).

The advantages of using GUS as a reporter gene are several. Most

GUS Protocols: Using the GUS Gene as a Reporter of Gene Expression

importantly, with few exceptions plants lack appreciable GUS activity. In spite of recent reports that many plants do in fact have low-level endogenous GUS-like activity with an acidic pH optimum, the background activity can usually be controlled by adequately buffering the assay pH to 8.0 or by including 20% methanol in the assay (Martin *et al.*, Chapter 2). And, even though animal tissue contains GUS, the pH optimum for mammalian enzyme (Gallie *et al.*, Chapter 13; Kyle *et al.*, Chapter 14) is much more acidic compared to GUS from *E. coli,* allowing the *E. coli* enzyme to be used as a reporter enzyme in animal cells as well. This is described in detail by Gallie *et al.* (Chapter 13). It is likely that in addition to GUS from *E. coli,* the mammalian enzyme will eventually be used as a reporter in several organisms including plants. Simply by assaying GUS activity at pH 4–5 and pH 7–8, it should be possible to measure two distinct reporter enzyme activities in the same tissue or tissue extract. In addition to plants and higher animals, other organisms that appear to be suitable for use with GUS (Jefferson, 1989) include most bacteria (Wilson *et al.*, Chapter 1), yeast, fungi (Schafer *et al.*, 1991), and many insects.

GUS assays are also straightforward, and substrates suitable for both histochemical, spectrophotometric, and fluorometric analysis (Naleway, Chapter 4) are readily available from a variety of companies (see appendix). The most popular in solution assay uses fluorescence (Jefferson, 1987). The fluorometric GUS assay based on the MUG substrate (Figure 1). A exercise demonstrating GUS enzyme kinetics using MUG is given by Gallagher (Chapter 3) and is suitable for an intermediate or advanced biochemistry teaching laboratory.

The 7-hydroxy-4-methylcoumarin (MU) is maximally fluorescent when the phenolic hydroxyl is ionized and as a consequence the assays are stopped with basic buffer (sodium carbonate). The assay is safe and nonradioactive, simple to perform, linear over extended periods of time, extremely sensitive, and inexpensive fluorometers are available for quantitation of the fluorescence (Gallagher, Chapter 3; see appendix for a list of fluorometer manufacturers). Luciferase, for example, requires costly chemicals and equipment (a luminometer) for quantitation. And even though luciferase is a popular reporter in many systems, the stability of the enzyme can be a problem. In Chinese hamster ovary cells, for example, the half-life of luciferase is 40 min (Gallie *et al.*, Chapter 13). GUS is very stable at 37°C used to culture animal cells and is highly resistant to high-temperature denaturation. Prior to histochemical staining, endogenous background activity in spruce or pine can be minimized by first heating the tissue to 60°C for 10 min (Stomp, Chapter 7). GUS stability can actually be a disadvantage because the enzyme

4-Methylumbelliferyl β-D-glucuronide (MUG) $\xrightarrow{\text{GUS}}$ glucuronic acid + 7-hydroxy-4-methylcoumarin (MU).

Fig. 1

accumulates over time and low levels of expression may appear artificially high, requiring verification of the results at the RNA level through blotting and probing (Martin *et al.,* Chapter 2).

GUS activity is also resistant to fixation, which improves visible light and transmission electron microscopic histochemical localization. However, depending on the sample and resolution required for analysis certain precautions must be observed. These are outlined in detail by Martin *et al.* (Chapter 2), Craig (Chapter 8), Stomp (Chapter 7), Finnegan (Chapter 11), and Kyle *et al.* (Chapter 14). Traditional GUS assays are destructive and require either slicing tissue for histochemical analysis or homogenization for measurement of activity in extracts. Nondestructive assays of the whole plant using both X-Gluc and MUG are also possible with GUS and are introduced by Martin *et al.* (Chapter 2).

In some situations, such as large-scale screening of transformed plant tissue for GUS activity, so many samples need to be analyzed that conventional methods and instrumentation do not work efficiently. Ideally, the whole process from tissue homogenization to enzyme assay quantitation should be automated. Brumback (Chapter 5) looks at the use of robotics to automate large-scale plant tissue processing to generate a protein extract for later enzyme assay. The system described is capable of grinding and extracting 96 samples in 45 min or 768 in an 8-h day. With the addition of microtiter plate-reading fluorometers (Rao and Flynn, Chapter 6), 96 GUS assays can be quickly quantitated at one time to further speed analysis.

Transformation with the GUS gene is possible through a variety of physical and biological techniques. *Agrobacterium* (Finnegan, Chapter 11), electroporation (Gallie *et al.,* Chapter 13; Osbourn and Wilson, Chapter 10), and particle gun (Stomp, Chapter 7 and cover photograph; Bowen, Chapter 12) mediated transformation are all illustrated in this book. The particle gun technique (e.g., Klein *et al.,* 1988), also known as "biolistics," shoots small particles of DNA coated tungsten or gold into tissue, thus sidestepping the need to create protoplasts prior to introducing the transforming DNA. In plant research, the particle gun is a key component in developing genetically engineered crops, and ease of histochemically locating the blue spots of GUS activity representing the transformed cells around each impact area (see cover photograph) has greatly aided analysis. Alternative reporter enzymes are, however, sometimes needed. Anthocyanin markers, for example, are particularly suited for particle gun research. Discussed by Bowen (Chapter 12), these markers permit cell-autonomous visualization of anthocyanin gene expression in living plants without tissue destruction.

The information in this book—references, suppliers, techniques, protocols, and applications—will serve as an important reference source for both researchers and students alike. It is particularly suited for classes in plant and animal molecular biology, biochemistry, and cell biology.

References

Jefferson, R. A. (1987). Assaying chimeric genes in plants: The GUS gene fusion system. *Plant Molec. Biol. Rep.* 5:387–405.

Jefferson, R. A., Burgess, S. M., and Hirsh, D. (1986). β-Glucuronidase from *Escherichia coli* as a gene-fusion marker. *PNAS* 83:8447–8451.

Jefferson, R. A., Kavanagh, T. A., and Bevan, M. W. (1987). GUS fusions: β-Glucuronidase as a sensitive and versatile gene fusion marker in higher plants. *EMBO J.* 6:3901–3907.

Klein, T. M., Gradziel, T., Fromm, M. E., and Sanford, J. C. (1988). Factors influencing gene delivery into *Zea mays* cells by high-velocity microprojectiles. *Bio/Technology* 6:559–563.

Koning, A., Jones, A., Fillatti, J. J., Comai, L, and Lassner, M. W. (1991). Arrest of embryo development in *Brassica napus* mediated by modified *Pseudomonas aeruginosa* exotoxin A. *Plant Mol. Biol.*, in press.

Schäfer, W., Stahl, D., and Mönke, E. (1991). Identification of fungal genes involved in plant pathogenesis and host range. In "Advances in Plant Gene Research," Vol. 8, "Genes Involved in Plant Defence" (F. Meins and T. Boller, eds.), Springer Verlag Wien, New York.

Summerfelt, K. R., Sheehy, R. E., and Hiatt, W. R. (1991). Recovery of transformed tomato plants without selection. 10th Annual Midwest Crown Gall Meeting. Purdue University, Lafayette, Indiana.

PART 1

The GUS Reporter Gene System

1 The *Escherichia coli gus* Operon: Induction and Expression of the *gus* Operon in *E. coli* and the Occurrence and Use of GUS in Other Bacteria

Kate J. Wilson[1]
Wye College,
University of London,
Wye, Ashford,
Kent, United Kingdom

Richard A. Jefferson[1]
Joint Division of the Food and Agriculture Organization of the United Nations and the International Atomic Energy Agency,
A1400 Vienna, Austria

Stephen G. Hughes[2]
Nuovo Crai
Caserta, Italy

The gene encoding β-glucuronidase, *gusA* (formerly *uidA*), which is now widely used as a reporter gene in plants and other organisms, was originally isolated from *Escherichia coli* (Jefferson *et al.,* 1986). In *E. coli, gusA* forms part of an operon. There are two genes downstream of *gusA,* one of which, *gusB,* encodes a glucuronide-specific permease; the function of the product of the third gene, *gusC,* is presently unknown. Upstream of *gusA,* and separately transcribed, is a gene, *gusR,* encoding a specific repressor of the *gus* operon. The primary focus of this chapter is to review the structure and functioning of the *gus* operon in *E. coli,* and to provide a protocol for the induction of the operon and

[1] Present address: Center for the Application of Molecular Biology to International Agriculture (CAMBIA), CAMBIA Organizational Office, Lawickse Allee 22, 6707 AG, Wageningen, The Netherlands.

[2] Present address: Plant Breeding International, Maris Lane, Trumpington, Cambridge, CB2 2LQ, United Kingdom.

assay of its products. We also discuss the rare occurrence of other GUS$^+$ bacteria, and how the contribution of these to "background" GUS activity can be both assayed and prevented. In addition, current uses of GUS as a reporter gene in bacteria are reviewed.

Occurrence and Natural History of Bacterial β-Glucuronidase

One of the key features that has led to the widespread adoption of GUS as a reporter gene in plant molecular biology is the absence of background activity in higher plants (Jefferson *et al.,* 1986, 1987). GUS activity is also restricted among many other groups of organisms, including bacteria. In fact, GUS assays are routinely used as diagnostic assays for the specific detection of *E. coli* and *Shigella* species in clinical and environmental samples (e.g., Rice *et al.,* 1990; Cleuziat and Robert-Baudouy, 1990). There were earlier reports of some GUS$^+$ *Salmonella* species (Killian and Buellow, 1979) but these appear not to have been confirmed by later studies (e.g., Perez *et al.,* 1986; Cleuziat and Robert-Bauduoy, 1990). The eighth edition of "Bergey's Manual" states that "it is taxonomically difficult to justify separate genera or even separate species status for (*E. coli* and *Shigella*)" (Brenner, 1984). Indeed *E. coli* and *Shigella* are serologically related and do exchange genetic information via intergeneric conjugation. Thus, at least among the Enterobacteriaceae, GUS activity can reasonably be said to be restricted to a single taxonomic group—that of *E. coli* and *Shigella* species.

The natural habitat of *E. coli* is the gut, and the GUS activity of *E. coli* plays a specific and very important role in its natural history. In vertebrates, one of the major pathways of detoxification of endogenous and xenobiotic organic compounds is by conjugation of these aglycones to glucuronic acid, a reaction carried out prominently in the liver, among other organs. The addition of the glucuronic acid group renders such hydrophobic compounds more water soluble, and enables them to be excreted in the bile or the urine (Dutton, 1966, 1980). Thus, the gut is a rich source of glucuronic acid compounds, providing a carbon source that can be efficiently exploited by *E. coli*. Glucuronide substrates are taken up by *E. coli* via a specific transporter, the glucuronide permease (see below), cleaved by *β*-glucuronidase, and the glucuronic acid residue thus released is used as a carbon source.

In general, the aglycone component of the glucuronide substrate is not used by *E. coli* and passes back across the bacterial membrane into the gut and is reabsorbed into the bloodstream. This circulation of hydrophobic compounds resulting from the opposing processes of glucuronidation in the liver and deglucuronidation in the gut is termed enterohepatic circulation (Figure 1). This phenomenon is of great physiological importance because it means that, due in large part to the action of microbial β-glucuronidase, many compounds, including endogenous steroid hormones and exogenously administered drugs, are not eliminated from the body all at once. Rather, the levels of these compounds in the bloodstream oscillate due to this circulatory process. This process is of great significance in determining pharmaceutical dosages,

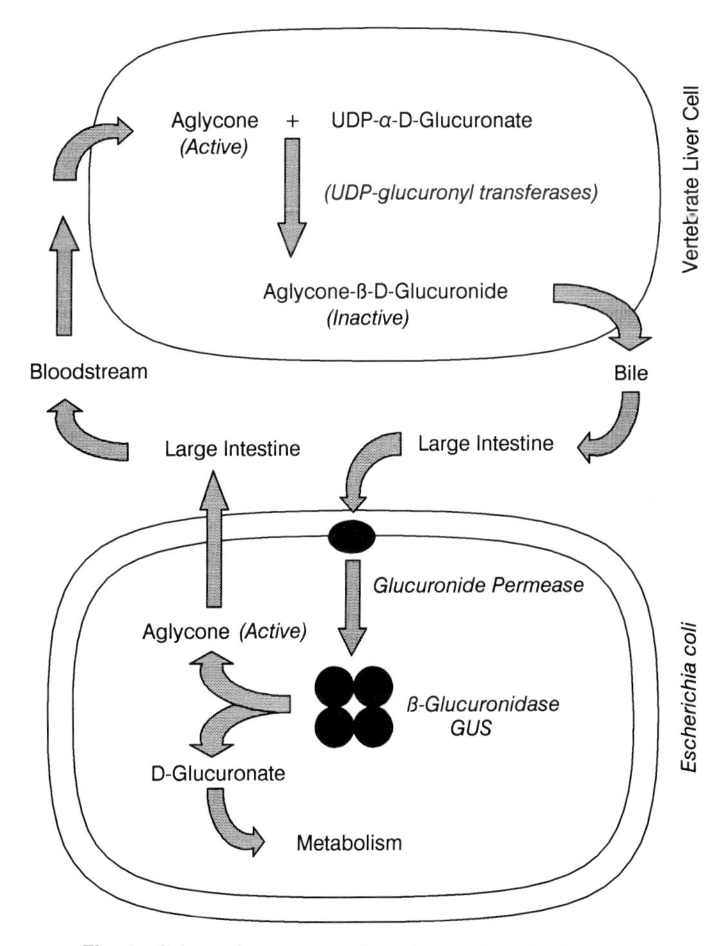

Fig. 1 Schematic representation of enterohepatic circulation.

and indeed some drugs are specifically administered as the glucuronide conjugate, relying on the action of β-glucuronidase to release the active aglycone (Draser and Hill, 1974). Enterohepatic circulation is also important in the day-to-day physiological state of the body, probably being a prime cause of the physiological impact of variations in diet or in gut flora (Goldin, 1986).

GUS activity is found in certain other bacterial species. In particular, it is found in other, nonenterobacterial, anaerobic residents of the gut, primarily in *Bacteroides* and *Clostridium* species (Hawkesworth *et al.*, 1971). Although these species exhibit lower β-glucuronidase activity per cell than *E. coli,* they are approximately 100-fold more abundant in the gut, and hence it was suggested that they might make a more significant contribution overall to enterohepatic circulation. However, it is difficult to judge the relative contributions of the different groups of bacteria based on a single set of measurements of their GUS activity with one glucuronide substrate. It is not known, for example, whether these organisms possess a glucuronide permease and whether their GUS activity, or any permease activity, possesses the same substrate versatility as those of *E. coli.*

There are reports of GUS activity in strains of *Streptococcus, Staphylococcus,* and *Corynebacteria* (Dutton, 1966), and we have found certain bacteria associated with plants that are GUS^+ (see below). However, GUS activity is not found in most of the bacterial species that are commonly studied because of their importance in agriculture, such as *Rhizobium, Bradyrhizobium, Agrobacterium,* and *Pseudomonas* species. Thus GUS is now being used as a reporter gene in these organisms, allowing studies of the spatial localization of gene activity of these bacteria in association with their plant hosts (see below).

The *gus* Operon in *E. coli*

The gene encoding β-glucuronidase, *gusA* (formerly *uidA*), maps at minute 36 on the *E. coli* chromosome, between the loci *add* (adenine deaminase) and *manA* (mannose-6-phosphate isomerase) (Novel and Novel, 1973). It has become clear that other genes involved in glucuronide metabolism and in regulation of β-glucuronidase activity map to the same region of the *E. coli* chromosome, forming the *gus* operon. The region has been studied extensively at a genetic and molecular-genetic level, and our current working model of the structure and functioning of the *gus* operon is summarized in Figure 2.

Regulation of GUS Activity in E. coli

β-Glucuronidase activity is not constitutively expressed in *E. coli:* rather, there appear to be three different factors regulating transcription of the operon. The primary mechanism of control is induction by glucuronide substrates. GUS activity is almost undetectable in cells that have been grown in the absence of glucuronides; however, incubation of *E. coli* in the presence of a glucuronide substrate leads to induction of high levels of GUS activity (Stoeber, 1961).

This regulation is due to the action of the product of the *gusR* (formerly *uidR*) gene, which encodes a repressor that is specific for the *gus* operon (Novel and Novel, 1976a). Inactivation or deletion of *gusR* leads to constitutive β-glucuronidase activity (Novel and Novel, 1976a; K. J. Wilson and R. A. Jefferson, unpublished results). The *gusR* gene maps to the same region of the chromosome as *gusA,* lying upstream of *gusA* and being separately transcribed. The direction of transcription is the same as *gusA* (Blanco *et al.,* 1985; K. J. Wilson and R. A. Jefferson, unpublished), and we have recently completed the DNA sequence of this region and have found that *gusR* is most likely encoded by an open reading frame of 195 amino acids.

GusR repression of β-glucuronidase activity has been shown by Northern analysis to be mediated by transcriptional regulation. RNA

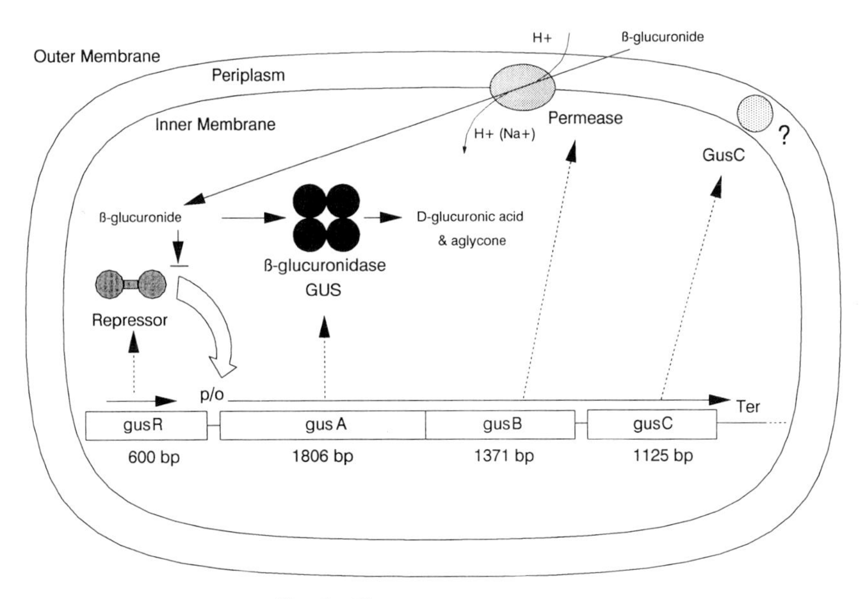

Fig. 2 The *gus* operon in *E. coli.*

from uninduced cultures of *E. coli* showed no hybridization to a *gusA* probe, in contrast to the strong hybridization observed to RNA extracted from cultures that had been induced with methyl β-D-glucuronide (Jefferson, 1985). Presumably, therefore, GusR acts by binding to *gusA* operator sequences so preventing transcription, this repression being relieved when a glucuronide substrate binds to the repressor and inactivates it. While the exact operator site remains to be defined, there are several candidate regions of dyad symmetry upstream of the *gusA* ATG (Jefferson *et al.*, 1986).

A second key level of control is that of catabolite repression. *Escherichia coli* grown in the presence of 1% glucose does not express β-glucuronidase activity even in the presence of a glucuronide inducer (Stoeber, 1961). A putative CAP binding site has been identified in the *gusA* upstream sequences (Jefferson *et al.*, 1986).

A third level of regulation of *gus* transcription appears to be exerted by the product of the *uxuR* gene. This gene, which maps elsewhere on the *E. coli* chromosome at minute 98, is primarily concerned with regulation of the *uxuAB* operon, which encodes enzymes involved in the further metabolism of glucuronic acid (Novel and Novel, 1976b). Repression of transcription of the *uxuA* and *B* genes is relieved by incubation with glucuronic acid, which presumably binds to and inactivates UxuR. In *E. coli* K-12, mutations in *uxuR* cause derepression of β-glucuronidase to only 1–4% of the full glucuronide-induced level (Novel and Novel, 1976b). As induction of transcription of a gene encoding an enzyme activity by the product of that enzyme activity is hard to understand, and as the level of regulation is <5% of that exerted by GusR and by CAP, it is not certain whether regulation by UxuR is of primary importance or is a secondary effect, perhaps resulting from some degree of homology between the two gene products. This latter possibility is supported by the observation that expression of *uxuR* on a multicopy plasmid can largely suppress the effect of a *gusR* mutation (Ritzenhaler *et al.*, 1983).

A Glucuronide-Specific Permease is Encoded by gusB

The existence of a glucuronide-specific permease was first demonstrated in the late 1950s by F. Stoeber working in the laboratory of Jacques Monod at the Pasteur Institute, Paris. Stoeber measured accumulation of [^{35}S]phenyl β-D-thioglucuronide, a glucuronide that is not hydrolyzed by GUS, in a fecal *E. coli* isolate and showed that this compound could be accumulated from external concentrations as low as 5 μM, giving over 200-fold concentration in the cell. This accumulation was inhibited in the presence of sodium azide, indicating that it is an active transport process. Using uptake competition studies he demonstrated that different glucuronides have different affinities for the permease, and some, such as phenolphthalein β-D-glucuronide, are not taken up at all. Like GUS activity, the permease activity is induced by preincubation with a glucuronide substrate (Stoeber, 1961).

Interestingly, Stoeber found that *E. coli* K-12, in contrast to the fecal strain, had very poor uptake ability (Stoeber, 1961). We have repeated this observation with the cloned *gus* operon from a K-12 strain and from a fecal strain. When expressed in a host strain that is deleted for the *gus* operon, the cloned K-12 genes conferred almost no ability to accumulate phenyl thioglucuronide in the cells, in contrast to the strong accumulation observed on expression of the operon from the fecal isolate (W.-J. Liang, P. J. Henderson, K. J. Wilson, and R. A. Jefferson, unpublished results). This observation is of considerable importance in induction assays (see below).

Molecular-genetic evidence concerning the glucuronide permease was first obtained when 340 bp of an open reading frame, encoding a highly hydrophobic protein, was identified immediately downstream of *gusA*—in fact there is a four base-pair overlap (ATGA) between the two genes (Jefferson *et al.*, 1986). The sequence of *gusB* from *E. coli* K-12 has now been completed and indicates a protein with 12 membrane-spanning domains that has 25% identity at the amino acid level to the *E. coli* melibiose transporter encoded by *melB* (R. A. Jefferson and W.-J. Liang, unpublished data; Liang, 1989). From the homology to the melibiose transporter (a sodium symporter) and from the dependence on membrane potential, it is likely that the glucuronide permease also functions as a cation symporter coupled to the electrochemical gradient.

The Range of *gus* Operon Inducers

We now know that efficient induction of expression of β-glucuronidase and of the whole operon depends on two key steps:

1. The substrate must be taken up into the cell via the glucuronide permease.

2. The substrate must be able to alleviate repression by the *gus* repressor.

Stoeber tested the ability of a number of different glucuronides to induce GUS activity, and found that it varied greatly, methyl β-D-glucuronide at 1 mM concentration inducing a level of GUS activity approximately 15 times that of phenyl β-D-thioglucuronide (a gratuitous inducer). He also found that some GUS substrates, such as phenolphthalein β-D-glucuronide and aldobiuronic acid (galactosyl β-D-glucuronide), do not act as inducers of β-glucuronidase activity (Stoeber, 1961). In the case of phenolphthalein β-D-glucuronide, this is clearly due to the inability of this substrate to be transported by the permease.

Using the protocol given below, we have tested the ability of GUS substrates now commonly used in quantitative and spatial analysis of GUS activity to act as inducers of the *gus* operon in *E. coli,* and have found that 5-bromo-4-chloro-3-indolyl β-D-glucuronide (X-Gluc), *p*-nitrophenyl β-D-glucuronide (PNPG), 4-methylumbelliferyl β-D-glucuronide (MUG), and resorufin glucuronide all act as powerful inducers. In general, values of GUS activity measured after 90 min of induction, starting with 1 mM external concentrations of these glucuronides, are of the order of 1–50 nmol PNPG hydrolyzed per minute per OD_{600} unit of bacterial culture. We have also tested a number of glucuronides that would occur naturally in the body, including estrogen glucuronide and testosterone glucuronide, and found that they too appear to have low inducing power (Liang, 1989).

This range of inducers illustrates a remarkable fact about the glucuronide permease, namely, that it is able to recognize and actively transport an extraordinary range of glucuronides with different aglycone residues. Likewise, the *gus* repressor must be able to recognize the same range of glucuronides. Thus, for example, while X-Gal does not act to induce the *lac* operon in *E. coli,* presumably because it does not relieve repression by the *lac* repressor and perhaps also because it is poorly transported by the *lac* permease, X-Gluc is a powerful inducer of the *gus* operon, indicating that it is actively accumulated within the cell by the glucuronide permease and acts to relieve the repression by GusR.

All the measurements discussed above, carried out both by Stoeber and by ourselves, have been on fecal isolates of *E. coli*. In contrast, *E. coli* K-12 strains show only very low levels of GUS induction in similar conditions. This can readily be visualized by streaking a fecal sample on an LB plate containing 50 μg/ml X-Gluc: overnight, dark blue colonies will appear. In contrast, an *E. coli* K-12 strain streaked on the same plate will show no or very little blue coloration in the colonies even after several days incubation. This is because the K-12 permease fails to concentrate the X-Gluc from the low external concentration in the plate (about 120 μM) and thus fails to induce the K-12 β-glucuronidase activity. To obtain high levels of K-12 β-glucuronidase induction it is necessary to use external concentrations of inducer of at least 10 mM.

Other Bacteria Show Inducible GUS Activity: Possible Contribution to "Background" Activity in GUS Assays

As discussed above, there are bacterial species other than *E. coli* that possess β-glucuronidase activity, particularly among the gram-positive genera *Staphylococcus* and *Streptococcus*. While the natural habitat of *Staphylococcus* is the skin and mucosal membranes of mammals, and that of *Streptococcus* is the intestinal tract and other organs also of mammals, it is important to remember that such microorganisms are not restricted in their distribution to these habitats but can be found elsewhere in stable and transient niches.

Solid rich media containing X-Gluc (50 μg/ml) provides a convenient means of screening for GUS-positive microorganisms. We have screened samples of soil, fecal matter, and some plant tissue in this way, and all, depending on provenance, have been shown to contain them. Here we give four examples of noncoliform GUS^+ bacteria that we have found in association with plants and that are illustrative of the false positive results that can occur in carrying out GUS assays with putatively transgenic plant material.

The first bacterium was found among the organisms that grew when pollen was shaken from the flowers of greenhouse-grown tobacco directly onto X-Gluc plates. It had the characteristic colonial morphology associated with highly mobile or gliding bacteria. At 30°C fingerlike projections advanced from the point of inoculation at the rate of about 2 cm per day. The mobile circular mound of cells at each end of the projection exhibited GUS (or GUS-like) activity. Of the recognized gliding bacteria, this organism most closely resembles the Myxobacteria. However, although spores were observed within the mounds, we did not observe the larger, sculpted fruiting bodies often found in this group. Of particular interest is the observation that when pollen grains of tobacco placed on X-Gluc plates are engulfed by these bacterial swarms, an intense deposition of blue pigment characteristic of a GUS X-Gluc reaction appears within the grains (Color Plate 1). Whether or not this observation represents genuine GUS activity, the phenomenon clearly illustrates the capacity of microorganisms to confound histochemical procedures, and emphasizes the need for caution. It is certainly possible that some of the observations of GUS or GUS-like activity in control plants could be explained by the presence of this type of organism.

The second and third organisms were both isolated directly from plant tissues that had been subjected to particle bombardment. More specifically, they were recovered from blue zones that developed around the points of impact of shrapnel following histochemical staining with X-Gluc. Fatty acid profiles did not exactly match any characterized microorganisms, but one showed similarity to a *Micrococcus* or *Clavibacter* species and the second to a *Staphylococcus* or *Brevibacterium* species. These assignments are consistent with the results of fermentation and enzyme activity profiles. Both organisms could be cultured on the surface of potato tubers and gave blue zones when the slices were subjected to histochemical staining with X-Gluc.

The fourth organism was isolated from yam (*Dioscorea cayanenesis*), again from tissue that had been subjected to particle bombardment. Analysis of the fatty acid content of this organism identified it as a *Curtobacterium* species, a genus of bacteria that is commonly isolated from plants. The GUS activity of this bacterium was studied more carefully and it was found not to be constitutive but to be inducible by incubation with each of the GUS substrates X-Gluc, MUG, and PNPG— albeit to levels below those observed in *E. coli* (M. Tor and K. J. Wilson, unpublished results). This induction of GUS activity was completely repressible by inclusion of 100 μg/ml chloramphenicol in the medium.

The studies on the latter microorganism indicate that GUS activity is inducible not just in *E. coli* but also in other microorganisms. Thus, some of the "background" activity that is observed when samples of tissue are subjected to long incubation in GUS substrates—often overnight—may result from the induction and subsequent activity of GUS in plant-associated microorganisms. If the enzyme is inducible rather than constitutive then it is easy to overcome the problem by the inclusion of chloramphenicol in the incubation medium, as this specifically inhibits bacterial, but not eukaryotic, protein synthesis. Another approach that has recently been demonstrated to effectively suppress "endogenous" GUS activity in plants is the addition of 20% methanol to the assay buffer (Kosugi *et al.*, 1990; see also Martin *et al.*, Chapter 2, and Stomp, Chapter 7). It is possible that, in some cases, the background activity being suppressed is due to GUS^{+} microorganisms.

Use of GUS as a Reporter Gene in Plant-Associated Bacteria

The use of GUS as a reporter gene in plant-associated bacteria has lagged far behind its use as a reporter gene in plants. Recently, however, its potential has begun to be realized.

Sharma and Signer constructed *gus* transposons based on transposon Tn5 that create either transcriptional (Tn*5*-*gusA*1) or translational (Tn*5*-*gusA*2) fusions to genes adjacent to the site of insertion. They used these transposons to create *gus* fusions to *Rhizobium meliloti* genes required for nodulation (*nod*) and symbiotic nitrogen fixation (*nif* and *fix*), and demonstrated the different spatial patterns of expression of these two classes of genes within the legume root nodule (Sharma and Signer, 1990). A similar promotor-probe *gus* transposon based on the Tn*3*-HoHo *lacZ* transposon has been constructed and is being used to analyze *hrp* genes of *Pseudomonas syringeae* (D. Dahlbeck, R. Innes, and C. Boucher, personal communication).

GUS fusions are also being used to look at targeting of viral and microbial proteins within plant cells. Translational fusions to *gusA* that added the entire coding region of the tobacco etch potyvirus proteins Nla and Nlb (encoding respectively a 49-kDa proteinase and a 58-kDa RNA polymerase) were shown to target the hybrid GUS Nla/Nlb proteins to the nucleus of the plant cell, the location in which the native Nla and Nlb proteins are normally found. This system can now be used as a powerful means of identifying the signals within the Nla and Nlb protein sequences that target them to the nucleus (Restrepo *et al.*, 1990). A similar approach using GUS fusions is being taken to study targeting of the VirD2 protein of *Agrobacterium tumefaciens* within the plant cell (J. Zupan and P. Zambryski, personal communication).

Finally, we are developing the *gus* operon as a transgenic marker system for the detection and monitoring of bacteria in soil and in association with plants. In initial experiments expression of the *gus* operon has been demonstrated in *R. meliloti* and used to detect the presence of marked strains in alfalfa root nodules by simply infiltrating the roots with buffer containing X-Gluc and observing the development of blue-coloured root nodules (Color Plate 2). Likewise *gusA* expression in a

Bradyrhizobium strain has been used to detect infection threads and to visualize areas of the root surface densely colonized by marked bacteria (Wilson *et al.*, 1991).

Protocol for Induction and Assay of E. coli *β-Glucuronidase*

The following protocol can be used to examine the induction and expression of β-glucuronidase activity in *E. coli* or in any GUS^+ bacteria that might be found in association with plant material. The GUS assay described here can of course be used to measure GUS activity in any bacterial strain, including that of GUS fusions constructed to analyze the regulation of specific bacterial genes.

Reagents

Minimal medium for growth of bacteria, using glycerol or succinate as carbon source.

Chloramphenicol (Cm)10 mg/ml in methanol
GUS assay buffer:
50 m*M* $NaPO_4$, pH 7.0
5 m*M* DTT
1 m*M* EDTA
100 m*M* PNPG stock, dissolve PNPG at 35 mg/ml in water
100 m*M* stocks of substances to be assayed for inducing power, [e.g., 100 m*M* MUG: dissolve at 39 mg/ml in dimethylformamide; 100 m*M* X-Gluc: dissolve at 40 mg/ml in dimethylformamide (Na^+ salt) or dissolve at 52 mg/ml (cyclohexammonium salt)]
Chloroform
0.1% SDS
0.4 *M* Na_2CO_3

Induction of Strains

1. Grow overnight cultures of strains in minimal medium. The minimal medium used (e.g., M9 or M63) (Miller, 1972) should first be checked and modified as necessary with the addition of specific growth requirements to ensure that the strains do grow! Glycerol

or succinate should be used as the carbon source, not glucose, to avoid any effects of catabolite repression.

2. Subculture strains into 1 ml minimal medium plus appropriate antibiotics and grow for at least 1 h to reach exponential phase.
3. Add 10 μl of 100 m*M* inducer. Keep one tube as a negative control without inducer. Grow for 1–3 h.
4. Transfer approximately 1.5 ml cells to an Eppendorf tube, spin 30 s, pour off supernatant and wash pellet by resuspending in 1 ml 1 × M9 salts + 100 μg/ml Cm. The washing removes all hydrolyzed glucuronide product, and the chloramphenicol prevents further protein synthesis. Centrifuge again, and resuspend pellet in 1.5 ml 1 × M9 salts + 100 μg/ml Cm.
5. Measure the OD of 0.5 ml at 600 nm. It is better if it is 0.5 or less. Keep remaining cells on ice or freeze at −70°C until ready for GUS assays.

GUS Assays

1. Thaw tubes on ice if necessary.
2. Prepare GUS assay buffer with 1.25 m*M* PNPG (add 12.5 μl 100 m*M* PNPG/ml). Prewarm to 37°C.
3. Permeabilize cells by vortexing with 1 drop 0.1% SDS and 2 drops chloroform for 10 s.
4. Prepare one Eppendorf tube per sample containing 800 μl GUS assay buffer with 1.25 m*M* PNPG. Add 200 μl permeabilized cell suspension (take care to avoid the chloroform which will be at the bottom of the tube). This gives a final concentration of PNPG of 1m*M*. Note approximate time.
5. Place reactions at 37°C. Keep an eye on the development of yellow color (*p*-nitrophenol). At at least three time points remove 100 μl into 800 μl 0.4 *M* Na_2CO_3. The time points do not have to be equally spaced, but note each time.
6. Measure absorbance at 405 nm for each time point against a substrate blank or a stopped blank reaction. Under these conditions the molar extinction coefficient of *p*-nitrophenol is 18,000. Thus in the 0.9 ml final volume, an absorbance of 0.020 represents 1 nmol of product produced.
7. Calculate the rate of the reaction in nanomoles product per minute as outlined below. The use of several time points enables the rate of the reaction to be calculated from the linear enzyme kinetics (see Jefferson and Wilson, 1991, for further discussion). Alternatively, a single measurement could be taken at a given

time after the start of the reaction, as is normally done for β-galactosidase (Miller, 1972). Normalize to OD units, or, if preferred, to cell protein or to viable cell number.

Calculation

For each sample plot a graph of OD_{405} (Y-axis) versus time in minutes (X-axis). Calculate the slope S of the graph (which should be linear!) in OD_{405} units per minute.

Rate of reaction R in nanomoles product per minute per OD_{600} unit is then

$$R = S/(0.02 \times V \times OD_{600})$$

where V is the volume assayed in milliliters. In the protocol listed above, $V = 0.02$ ml because 0.2 ml sample is used in the initial reaction (step 4) and one-tenth of this (100 μl) is removed into stop buffer for each time point (step 5).

This measurement differs from the Miller units used to measure β-galactosidase in that Miller units are calculated as OD_{420} units $\times$ 1000 per min per ml of culture at OD_{600} measured in a final reaction volume of 1.7 ml; that is, they are not converted to actual nanomoles product produced per unit time per unit of culture (Miller, 1972).

Note: This assay can very easily be automated using commercially available ELISA plate readers and microtiter equipment. The calculation given above derives from the Beer-Lambert law and assumes a pathlength of 1 cm, as found in most spectrophotometers. However, if a microtiter dish and an ELISA reader are used, the path length will be different and the conversion factor must be recalculated. This can be done using commercially available liquid *p*-nitrophenol as a standard to calibrate the readings for specific volumes of solution assayed in the microtiter wells. Absorbance can also be measured at 415 nm, at which wavelength the molar extinction coefficient of *p*-nitrophenol is 14,000.

Acknowledgments

We thank Mahmut Tor, Sinclair Mantell, and Charles Ainsworth for sharing unpublished results.

References

Blanco, C., Mata-Gilsinger, M., and Ritzenhaler, P. (1985). The use of gene fusions to study the expression of *uidR,* a negative regulatory gene in *Escherichia coli* K-12. *Gene* 36:159–167.

Brenner, D. J. (1984). Enterobacteriaceae. In "Bergey's Manual of Systematic Bacteriology" (N. R. Krieg, ed.), Vol 1, pp. 408–423. Williams and Wilkins, Baltimore.

Cleuziat, P., and Robert-Baudouy, J. (1990). Specific detection of *Escherichia coli* and *Shigella* species using fragments of genes coding for β-glucuronidase. *FEMS Microbiol. Lett.* 72:315–322.

Draser, B. S., and Hill, M. J. (1974). "Human Intestinal Flora." Academic Press, New York.

Dutton, G. J., ed. (1966). "Glucuronic Acid, Free and Combined." Academic Press, New York.

Dutton, G. J. (1980). "Glucuronidation of Drugs and Other Compounds." CRC Press, Boca Raton, Florida.

Goldin, B. R. (1986). In situ bacterial metabolism and colon mutagens. *Annu. Rev. Microbiol.* 40:367–393.

Hawkesworth, G., Draser, B. S., and Hill, M.J. (1971). Intestinal bacteria and the hydrolysis of glycosidic bonds. *J. Med. Microbiol.* 4:451–459.

Jefferson, R. A., (1985). DNA Transformation of *Caenorhabditis elegans:* Development and application of a new gene fusion system. Ph.D. Dissertation, University of Colorado, Boulder.

Jefferson, R. A., and Wilson, K. J. (1991). The GUS gene fusion system. In "Plant Molecular Biology Manual" (S. Gelvin, R. Schilperoort, and D. P. Verma, eds.), B14, pp. 1–33, Kluwer Academic Publishers, The Netherlands.

Jefferson, R. A., Burgess, S. M., and Hirsh, D. (1986). β-Glucuronidase from *Escherichia coli* as a gene fusion marker. *Proc. Natl. Acad. Sci. USA* 86:8447–8451.

Jefferson, R. A., Kavanagh, T. A., and Bevan, M. W. (1987). GUS fusions: β-glucuronidase as a sensitive and versatile gene fusion marker in higher plants. *EMBO J.* 6:3901–3907.

Kilian, M. and Buellow, P. (1979). Rapid identification of Enterobacteriaceae. II. Use of a β-glucuronidase detecting agar medium (PGUA agar) for the identification of *E. coli* in primary cultures of urine samples. *Acta Pathol. Microbiol. Scand. B* 87:271–276.

Kosugi, S., Ohashi, Y., Nakajima, K., and Arai, Y. (1990). An improved assay for β-glucuronidase in transformed cells: Methanol almost completely suppresses a putative endogenous β-glucuronidase activity. *Plant Sci.* 70:133–140.

Liang, W.-J. (1989). Studies on the glucuronide operon of *Escherichia coli.* M.Sc. Thesis, Cambridge University, UK.

Miller, J. H. (1972). "Experiments in Molecular Genetics." Cold Spring Harbor Laboratory, Cold Spring Harbor, NY.

Novel, G., and Novel, M. (1973). Mutants d'*Escherichia coli* affectes pour leur croissance sur methyl β-glucuronide: Localisation du gene de structure de la β-glucuronidase (*uidA*). *Mol. Gen. Genet.* 120:319–335.

Novel, M., and Novel, G. (1976a). Regulation of β-glucuronidase synthesis in *Escherichia coli* K-12: Constitutive mutations specifically derepressed for *uidA* expression. *J. Bacteriol.* 127:406–417.

Novel, M., and Novel, G. (1976b). Regulation of β-glucuronidase synthesis in *Escherichia coli* K-12: Pleiotropic constitutive mutations affecting *uxu* and *uidA* expression. *J. Bacteriol.* 127:418–432.

Perez, J. L., Berrocal, C. I., and Berrocal, L. (1986). Evaluation of a commercial β-glucuronidase test for the rapid and economical identification of *Escherichia coli. J. Appl. Bacteriol.* 61:541–545.

Restrepo, M. A., Freed, D. D., and Carrington, J. C. (1990). Nuclear transport of plant potyviral proteins. *Plant Cell* 2:987–998.

Rice E. W., Allen, M. J., and Edberg, S. C. (1990). Efficacy of β-glucuronidase assay for the identification of *Escherichia coli* by the defined substrate technology. *Appl. Environ. Microbiol.* 56:1203–1205.

Ritzenhaler, P., Blanco, C., and Mata-Gilsinger, M. (1983). Interchangeability of repressors for the control of the *uxu* and *uid* operons in *E. coli* K-12. *Mol. Gen. Genet.* 191:263–270.

Sharma, S. B., and Signer, E. R. (1990). Temporal and spatial regulation of the symbiotic genes of *Rhizobium meliloti* in planta revealed by transposon Tn*5*-*gusA*. *Genes Dev.* 4:344–356.

Stoeber, F. (1961). Etudes des proprietes et de la biosynthese de la glucuronidase et de la glucuronide-permease chez *Escherichia coli.* These de Docteur es-Sciences, Paris.

Wilson, K. J., Giller, K. E., and Jefferson, R. A. (1991). β-Glucuronidase (GUS) operon fusions as a tool for studying plant-microbe interactions. In "Advances in Molecular Genetics of Plant–Microbe Interactions" (H. Hennecke and D. P. S. Verma, eds.), Vol. 1, pp. 226–229. Kluwer Academic Publishers, The Netherlands.

2 The GUS Reporter System as a Tool to Study Plant Gene Expression

Thomas Martin, Rosa-Valentina Wöhner, Sabine Hummel, Lothar Willmitzer, and Wolf B. Frommer
Institut für Genbiologische Forschung
Berlin, Federal Republic of Germany

During the last few years the bacterial β-glucuronidase gene (*uidA, gusA*), commonly referred to as the GUS gene, combined with the increasing number of plant species accessible to molecular transformation, has become the major reporter gene used as a tool for the analysis of plant gene expression (Willmitzer, 1988; Walden and Schell, 1990, Table 2). The wide acceptance is mainly due to such advantages as fast and nonradioactive analysis. The assay is extremely sensitive, and it is possible to obtain both quantitative (i.e., level of expression) and qualitative (i.e., specificity of expression in tissues and organs) data with the same reporter gene. Quantitative assays are performed using fluorigenic substrates such as 4-MUG (4-methylumbelliferryl-β-glucuronide), whereas X-gluc (5-bromo-4-chloro-3-indolyl-β-D-glucuronide), on the other hand, can qualitatively show cell- and tissue-specificity. Chemical modification of the substrate might even allow use of GUS as a selectable marker (Jefferson, 1989).

GUS is used for a wide range of applications. We and others have found promoter–GUS fusions very useful for promoter analysis and for dissecting gene families (Stockhaus *et al.,* 1989; Keil *et al.,* 1989; Rocha-Sosa *et al.,* 1989; Köster-Töpfer *et al.,* 1989, 1990; Liu *et al.,* 1990, 1991). For this purpose Jefferson (1987) has developed a set of vectors that are based on the binary plasmid pBIN 19 (Bevan, 1984) and

GUS Protocols: Using the GUS Gene as a Reporter of Gene Expression

that allow both transcriptional fusions and translational fusions in all three reading frames. Among these, a derivative containing an improved ATG that follows the Kozak rule (Kozak, 1989) and a plasmid containing the CaMV 35S promoter fused to GUS are also available. Furthermore a multiplicity of derivatives in pUC-based vectors have been constructed (e.g., Töpfer *et al.,* 1988).

GUS has also proven to be useful for protein targeting studies, for example, in mitochondria (Schmitz *et al.,* 1990) and to develop and optimize transformation protocols for higher plants. This is especially true for monocotyledonous plants and fungi (Janssen and Gardner, 1989; Gould and Smith, 1989; Schäfer *et al.,* 1991). The introduction of a portable intron into the GUS gene leads to nearly complete repression of its expression in *Agrobacteria* (Vancanneyt *et al.,* 1990) and thus allows one to visualize transformation events of single cells at early stages after cocultivation. Furthermore, the GUS marker should be useful for studying plant–pathogen interactions as either native GUS activity or *gusA*-derived activity from transformed fungi can be used to follow the infection process in the host plant (Schäfer *et al.,* 1991).

The disadvantage of the GUS system, in comparison to the luciferase reporter system (Ow *et al.,* 1986; Olsson *et al.,* 1988), of involving destructive assays, has been removed by the introduction of three new nondestructive assay systems (Gould and Smith, 1989; Martin *et al.,* in press).

One important prerequisite for the use of reporter genes is the absence of endogenous activities in the organisms to be transformed. Jefferson *et al.* (1987) have analyzed for the presence of GUS activity and found no endogenous activity in a broad spectrum of higher plants, whereas Plegt and Bino (1989) found activity in the male gametophyte of several solanaceous plants. Hu *et al.* (1990) have undertaken an extensive study of 52 seed-plant species, and showed that plants usually lack activity in vegetative organs but show activity in reproductive tissues. The assays in all three sets of experiments were performed according to the standard protocol from Jefferson (1987). Unpublished observations from A. Alwen, O. Vicente, and E. Heberle-Bors have provided evidence for endogenous β-glucuronidase activity with a pH optimum of pH 5 to be present in a wide variety of species. Based on these observations we have examined *Arabidopsis thaliana* for the presence of endogenous GUS activity and found similar activities in both vegetative and floral organs at pH 5 (Color Plate 1) (T. Martin, R. V. Wöhner, W. B. Frommer, unpublished results).

Standard Assays

Here we describe a set of methods as commonly used for most applications in plant molecular biology, including some remarks and comments. The methods described in this section involve either homogenization of plant material in order to obtain quantitative values or involve dissection, sectioning, fixation, or destaining of the plants for histochemical staining, which involve destruction of the living material. Further details are given in Jefferson (1987). Two basic methods are available to detect GUS activity. The fluorescence assay allows quantitation of GUS activity in protein extracts, whereas the histochemical assay allows one to determine the cell and tissue specificity. See Naleway (Chapter 4) for more details on substrates.

The fluorescence assay is extremely sensitive for the detection of low levels of promoter activity, if the promoter is active in a large number of cells. In case of localized expression the histochemical assay is much more sensitive.

Standard Protocol for Quantitation of GUS Activity

Cleavage of the substrate 4-MUG by a β-glucuronidase activity leads to the generation of the fluorigenic product 4-MU, which can be visualized or detected by irradiation with UV light.

Basic GUS Assay

1. Use fresh material or material frozen in liquid nitrogen.
2. Extract 50–200 mg tissue in an Eppendorf tube with motor-driven homogenizer (e.g., Heidolph homogenizer RZR 50, Kehlheim, FRG) in 200–300 μl extraction buffer [50 m*M* Na-phosphate pH 7, 10 m*M* β-mercaptoethanol, 10 m*M* ethylenediamine tetraacetic acid (EDTA), 0.1% sodium dodecyl sulfate (SDS), 0.1% Triton X-100].
3. Centrifuge 15 min at 4°C at 18 krpm.
4. Take 200 μl of the supernatant.
5. Measure protein concentration according to, for example, Bradford (1976).

6. Add 20 μg (10–20 μl) protein extract to 200 μl 2 m*M* 4-MUG (Sigma, Heidelberg, F.R.G.).
7. Incubate duplicate samples for 0 (blank) and 30 to 60 min at 37°C.
8. Stop reaction by adding 0.8 ml 0.2 *M* Na_2CO_3.
9. View under UV/photograph with Wratten 2E filter (Kodak).

Quantitation of Enzyme Activity

10. Determine enzyme activity (see also Gallagher, Chapter 3): 4-MU standard solutions: dissolve 8.8 mg 4-MU in 5 ml dimethyl sulfoxide (DMSO), dilute this stock 1:100 in DMSO, further dilute 1:200 in 0.2 *M* Na_2CO_3.
11. Measure difference to blank at 365 nm excitation/456 nm emission in a fluorimeter (e.g., Hoefer TKO 100, Kontron SFM25, München, F.R.G.; see Gallagher, Chapter 3, for more details).
12. Calculation of the enzyme activity: activity (pmol/mg protein/min) = relative fluorescence (%) × 1000 p*M*/100% × mg protein × min.

Comments

Problems can arise from errors in determining the amount of protein, from protein degradation, from fluorigenic compounds present in the tissue, from endogenous activity of the untransformed plant, or from contamination by soil bacteria or fungi.

Problem 1: Analysis of Roots

We have found that the above-described homogenization protocol yields only low amounts of protein from roots. Probably due to interference by phenolic compounds, the determination of protein content in root extracts using the Bradford assay is not very reliable. Furthermore we have observed that samples from roots were occasionally degraded. It is therefore recommended to homogenize root tissue with mortar and pestle in the presence of fine sand and 0.2 m*M* PMSF. The problem of determination of the amount of protein can be circumvented by measuring the concentration according to Lowry *et al.* (1951) after trichloroacetic acid (TCA) precipitation. Quality and quantity of the extracts can be analyzed by separating the proteins by SDS polyacrylamide gel electrophoresis (PAGE) and subsequent staining with Coomassie blue.

Problem 2: Contaminating Activities

The presence of GUS-positive soil bacteria can mimic promoter activity in roots isolated from soil grown plants. By growing plants under sterile conditions, contamination can be avoided.

It has been observed that native or contaminating β-glucuronidase activities derive from enzymes that differ in their molecular weight from the *gusA*-derived activity. This difference can be used to differentiate between the two by a zymogram, that is, an *in situ* detection of GUS on a polyacrylamide gel. As GUS is stable under denaturing conditions, it can be separated by electrophoresis in SDS-containing gels (Jefferson, 1987). Nondenaturing separations are also possible and are described by Kyle *et al.* (Chapter 14).

Zymogram for GUS Detection

1. Incubate extracts for 10 min at room temperature in 2% SDS, 62 m*M* Tris-Cl , pH 6.8, 5% β-mercaptoethanol, 10% glycerol, 0.0002% bromophenol blue.
2. Separate samples at room temperature on a 10% SDS PAGE (Laemmli, 1970).
3. Equilibrate gel for 2 h in GUS extraction buffer at room temperature.
4. Incubate gel in cold 1 m*M* 4-MUG in GUS extraction buffer.
5. Visualize under UV.
6. Document on polaroid with Wratten 2E filter.

Problem 3: Fluorigenic Compounds

In order to control whether the detected activity is due to fluorigenic compounds other than 4-MU, a time kinetic can be performed. The GUS-derived fluorescence should increase with time, whereas the relative fluorescence due to other fluorigenic compounds should remain constant. Alternatively, substrates fluorescing at a different wavelength are available (e.g., resorufin-β-D-glucuronide (ReG), excitation 572 nm, emission 583 nm; see Naleway, Chapter 4]. Endogenous chromogenic compounds that mask or mimic GUS activity can be eliminated by the addition of insoluble polyvinylpyrrolidoue (PVP) to the material before homogenization or the extract can be passed over Sephadex G25 or G50 columns (Jefferson, 1987; Raineri *et al.*, 1990). The presence of a GUS activity can be verified by using the specific β-glucuronidase inhibitor saccharolactone (Jefferson, 1987).

A severe problem can be posed by the presence of endogenous GUS activities, especially in flowers and fruits (Plegt and Bino, 1989; Hu *et al.*, 1990). As described above, the endogenous activity is strictly pH dependent and can thus be controlled by shifting the pH of the extraction buffer to pH 8 (see also next section). It has recently been shown that organic solvents inhibit the endogenous GUS activity from solanaceous plants but enhance the *gusA*-dependent activity (Kosugi *et al.*, 1990). We have demonstrated that this holds true also for *Arabidopsis thaliana* (T. M., R. V. W., and W. B. F., see next section). The addition of 20% methanol to the reaction mixture drastically inhibits the plant-derived activity.

Standard Protocol for Histochemical GUS Assays

Analogous to the β-galactosidase substrate X-gal, which yields a blue color, the substrate X-gluc can be used to detect GUS activity. As the GUS enzyme is stable even after fixation, generally the tissue is fixed in paraformaldehyde prior to staining.

Fixation Protocol

1. Incubate hand sections, organs, or whole plants (e.g., *Arabidopsis* seedlings) for 30 min on ice in 2% paraformaldehyde, 100 m*M* Na-phosphate, pH 7, 1 m*M* EDTA.
2. Wash material in 100 m*M* Na-phosphate, pH 7.
3. Submerge material in 2 m*M* X-gluc (Biosynth, USA) in buffered solution (50 m*M* Na-phosphate, pH 7, 0.5% Triton X-100).
4. Vacuum infiltrate (approximately 20 mbar) for 10 s.
5. Incubate 10 min to 2 days at 37°C (check staining from time to time to avoid overstaining).
6. Stop by washing tissue in H_2O.
7. After staining green tissue must be bleached to visualize the blue staining; wash several times in 70% ethanol.
8. Stained material is stable at least for several months in ethanol (potato leaves, for example, are very fragile after destaining).

General Comments

In many cases the protocol has been simplified and the fixation is omitted. As the stainings are usually performed overnight, the incubation conditions might influence promoter activity. Important parameters that are known to influence the expression of different promoters are therefore light conditions, anaerobiosis due to submersion, osmolarity of the incubation solution, metabolizable buffer substances such as phosphate, or the influence of polyamines such as Tris buffers (Evans and Malmberg, 1989). This can lead to false interpretation of the results obtained. Examples for such promoters are the light-regulated photosynthesis genes (Tobin and Silverthorne, 1985), the anaerobically regulated maize Adh1 (Freeling, 1973), or the sucrose-regulated class I patatin (Rocha-Sosa *et al.,* 1989). To prevent the influence of metabolism on the staining during the incubation procedure, the material should either be fixed or stainings performed under different conditions (e.g., buffers, light, etc.) for comparison.

For the staining of flowers and fruits the addition of 10–100 m*M* ascorbate is recommended in order to prevent browning of the tissue.

In order to evaluate the important parameters of the different published variants of the protocols, we have performed experiments using untransformed *Arabidopsis thaliana* C24 and *Arabidopsis* transformed with a chimeric patatin/GUS gene, which gives expression in roots (patatin B33; Rocha-Sosa *et al.,* 1989, W. B. Frommer and R. Schmidt, unpublished results), and a construct that gives a similar expression pattern to the CaMV 35S promoter, that is, strong constitutive expression in all tissues (D4E, a deletion subclone of the patatin B33 promoter fused to the CaMV 35S enhancer; Liu *et al.,* 1990, W. B. Frommer and R. Schmidt, unpublished results). Plants were grown on 3MS medium (Murashige and Skoog, 1962) or in soil. Two sets of experiments were performed independently with three plants each.

Influence of Incubation Time and X-Gluc Concentration

The influence of incubation time and X-gluc concentration on the staining pattern of B33-GUS and D4E-GUS transformants was tested according to the standard protocol described above. The staining was monitored by eye every 15 min for the first 4 h and again after 16 h.

At a concentration of 0.1 m*M* X-gluc, in case of the constitutively expressed D4E-GUS plants, the first blue color was visible in roots after

45 min, in leaves after 75 min, in the stem after 210 min, and a maximum was reached after 4 h. For concentrations between 1 and 10 m*M*, the first staining was observed within 30 min in all organs and plateaued after 1 h. Whether the preferential staining of roots at low concentrations is due to higher accessability or to differential promoter activity was not further analyzed. Nevertheless, different results can be obtained depending on concentration of the substrate and length of incubation.

The B33-GUS plants showed the first visible activity in roots at 0.1 m*M* X-gluc after incubation overnight. At 1.0–10.0 m*M* X-gluc the first activity was visible in less than 1 h and the maximal staining was reached after 3 h.

Influence of Triton and Vacuum on X-Gluc Staining

For transformed plants 33-GUS and D4E-GUS, the role of Triton X-100 in the incubation solution and the role of vacuum treatment were analyzed. Vacuum treatment led to a reduction of the time necessary to detect the first visible staining. No effect in respect to intensity or pattern of the staining in roots and leaves could be found. The presence of 0.5% Triton X-100 gave a more uniform staining pattern in leaves, but showed no difference in the staining of roots.

Influence of Different Buffer Substances

As mentioned above, some buffer substances might have effects on certain promoters, if the tissues are not fixed before staining. Different buffers (phosphate, MES, Tris, citrate, or water, each adjusted to pH 7.0) at different concentrations in the staining solution showed no significant differences in both intensity and pattern of staining for both constructs.

Endogenous and Gus A-Derived β-Glucuronidase Activity in Dependence of the pH Value of the Staining Solution

Due to the strong native GUS activity present in the flowers and fruits of several plant species (Plegt and Bino, 1989; Hu *et al.*, 1990), histochemical analysis can be severely impaired. As the pH optimum of the endogenous β-glucuronidase activity of plants lies at pH 5 (A. Alwen, O. Vicente and E. Heberle-Bors, pers. comm.), we compared the staining patterns of wildtype *Arabidopsis thaliana* C24 and transformed lines B33-GUS and D4E-GUS after incubation in phosphate buffer ranging from pH 5 to pH 8. The pH of the staining solution strongly influences the endogenous activity of *A. thaliana* (Table 1). At pH values below pH 7, staining was detectable in leaves, stems, roots, and flowers of untransformed plants (Color Plate 3a,b). In leaves, the staining was preferentially in the vascular tissue. No endogenous activity could be detected at pH 7 and 8. GUS-transformed plants show *gusA*-derived activity at all pH values, but the activity is higher at values equal and above pH 7. Thus, by increasing the pH of staining solution, endogenous GUS activities can be suppressed. This could be especially useful to reduce the background activities present in flowers and fruits (Hu *et al.*, 1990).

In several protocols the buffer is omitted from the staining solution. As the deionized water often has an acidic pH value, the addition of a neutral or high-pH buffer is recommended in order to limit native GUS activities.

Reduction of Plant-Derived GUS Activity by Organic Solvents

As previously described by Kosugi *et al.* (1990), we find that the presence of 20% methanol in the staining solution clearly reduces the endogenous activity in all plant organs of wildtype plants. Neverthe-

Table 1

Dependence of the Expression of GUS on the pH Value and the Presence of Organic Solvents in the Incubation Solution

Arabidopsis/ transformant	Organ	pH 5	pH 6	pH 7	pH 8	pH 5, 20% MeOH
C24 (soil)	Anthers	+++	++	−	−	+
(wild-type)	Pistils	+++	++	−	−	−
	Seeds	+++	++	−	−	+
	Vegetative	nd	nd	nd	nd	nd
C24 (3MS)	Roots	++	+	−	−	+/−
(wild-type)	Stems	++	+	−	−	+/−
	Leaves	++	+	−	−	+/−
33-5 (3MS)	Roots	+++	++++	++++	++++	++++
D4E-h (3MS)	Roots	++++	++++	+++++	+++++	+++++
	Leaves	+++	+++	+++++	+++++	+++++

[a] Key: nd, Not determined; +, intensity of staining; −, not detectable with eyes; +/−, very weak activity.

less, low residual activity is detectable in anthers, seeds, and, at very low levels, also in roots, stems, and leaves. GUS transformed plants show no reduced activity in specific staining but an improved signal-to-background ratio by this method.

Influence of Oxidation Catalysts

To enhance oxidation of the reaction intermediate 5-bromo-4-chloro-indol to the blue product 5-bromo-4-chloro-indigo and thus exclude artifacts by local oxidative processes, a mixture of 0.5 m*M* ferricyanide, 0.5 m*M* ferrocyanide can be included in the incubation (Jefferson, 1987; see also Stomp, Chapter 7). Histochemical stainings of *Arabidopsis* seedlings transgenic for 33-GUS or D4E-GUS in the presence of cyanide are two to three times faster in respect to the first detectable color development. The staining pattern was similar in roots and leaves. However, after maximal staining plants incubated in the presence of cyanide show a more diffuse pattern when compared to plants stained in the absence of cyanide.

Influence of Fixation with Paraformaldehyde

We also compared the patterns and intensity of staining of nonfixed to fixed plantlets and found that fixation under the above-described conditions reduced the staining intensity and lead to a patchy pattern in roots and leaves in the case of both constructs. This could be due either to unequal fixation or local inhibition of the GUS enzyme.

Possible Disadvantages of the Stability of the GUS Enzyme

For evaluation of data obtained with β-glucuronidase one should notice the stability of the bacterial enzyme and thus the extreme sensitivity of the assay. Even very weak expression will eventually lead to significant staining and thus can lead to overinterpretation of the data. The advantage of being able to detect even low levels of expression can turn out to be a disadvantage if the developmental regulation of promoters is analyzed using the GUS reporter system. Due to the accumulation of GUS protein, it is difficult to differentiate between early onset of transcription and strong promoter activity. A marker gene more suitable for the analysis of differentiation seems to be the SPT gene (Jones *et al.*, 1989; see also Finnegan, Chapter 11).

GUS Determination on the RNA and Protein Level

In order to support the data obtained with the methods described above, Western blot analysis can be performed using the commercially available GUS-antibody preparations (for protocols, consult Sambrook *et al.*, 1989).

As the GUS protein is very stable, the protein accumulates over time. In order to confirm low levels of expression or to follow induction, or circadian rythms, the expression can be followed at the RNA level. RNA can be efficiently isolated from a wide variety of plant species and tissues according to Logemann *et al.* (1987), and Northern blot analysis can be performed according to Sambrook *et al.* (1989).

Nondestructive Assays

For several applications the above assays are not suitable as they involve destruction of the plant material. Destructive assays can be used to establish or improve transformation procedures by analyzing early transformation events, but in many cases the stained cells and their progeny are required to survive. Therefore nondestructive methods using 4-MUG as a substrate have been developed for the analysis of early transformation events (Gould and Smith, 1989). Also in all cases where an early analysis of seedlings is necessary, such as to test crossings or to follow the excision of transposable elements, a nondestructive assay system is essential.

The use of promoter GUS fusions has opened a new means of unraveling signal transduction pathways. A genetic approach should theoretically allow to identify all steps of the signal transduction chain. For this purpose a promoter regulated by the stimulus of interest is fused to a marker gene and the chimeric gene is transferred to *Arabidopsis*. *Arabidopsis* is used as it has several features which make the plant suitable for molecular genetics (Damm *et al.*, 1991). Seeds of transgenic plants can be mutagenized in order to obtain mutants in individual steps of the transduction chain, and the M2 generation is screened or selected for the altered phenotype. GUS could be the preferable marker for this screening, if the assay could be performed under nondestructive conditions, as it allows to differentiate between minor differences and as qualitative changes in the expression can be followed. This would allow one to visualize both quantitative and qualitative changes caused by the mutations without sacrificing the mutant. We have therefore developed three additional nondestructive protocols suitable for screening large populations of *Arabidopsis* plants.

Nondestructive Assay using MUG in Tissue Culture Media

In order to establish and optimize transformation protocols the detection of transformation events *in vivo* is useful. As 4-MUG does not seem to be toxic to petunia shoots during short incubation periods (up to 2 days), a nontoxic staining procedure in tissue culture media has been developed (Gould and Smith, 1989). Due to the leakage of β-glucuronidase from cultured plant tissues into the medium, GUS expression can be analyzed in the spent media after transfer of the material to new medium. Alternatively, protoplasts, calli, or suspension cultures can be stained directly without destruction of the material.

Nondestructive Assay Protocol

1. Culture material for 2 days in liquid or agar medium containing 2 m*M* 4-MUG.
2. Incubate overnight at 30°–37°C. The temperature depends on promoter strength.
3. Transfer to new medium.
4. Add 10–30 μl 0.3 *M* Na_2CO_3 to the tissue.
5. Evaluate staining after 20 min under UV light.

Comments

The assay can also be used to analyze the expression in young seedlings after introducing a small wound. This can be used, for example, for the analysis of offspring at early stages of development. Though the assay is nondestructive, it involves heat shock and in case of seedling analysis wounding of the plant and thus might influence the expression from certain promoters.

The above-described assay is restricted to genes that are expressed under the tissue culture conditions, which does not apply to all promoters. In order to be able to analyze intact plants with a minimum of manipulations, we have developed two new assay systems. Depending on the expression profile of the gene of interest, either roots or above ground parts of the plant can be analysed for GUS expression *in vivo*.

Nondestructive Assay using X-Gluc in Axenic Culture for Root Staining

As the blue product of the X-gluc staining is less mobile in the plant as compared to the fluorescent 4-MU, a staining procedure for roots using X-gluc was developed (Martin *et al.*, in press).

As in case of 4-MUG, X-gluc does not seem to be toxic if plants are exposed only for short periods. Transfer to new medium can thus be used to rescue the plants after staining (Color Plate 3c,d).

Nondestructive Assay Protocol

1. Grow seedlings axenically on MS medium in 15-cm petri dishes for ~10 days (two- to four-leaf stage).
2. Add 2 ml 2 m*M* X-gluc in H_2O (+0.5% Triton)(sterile filtrated).
3. Incubate overnight under normal growth chamber conditions (22°C).
4. Evaluate root staining.
5. Rescue by transfer to new MS medium.

Comments

Care should be taken when transfering the plantlets to new medium as plants are easily harmed at this stage. The roots can be better observed if plates are incubated vertically.

Nondestructive Assay by Spraying MUG

As the blue color from X-gluc staining is difficult to detect in green parts of the plant, a fluorescent assay suitable for GUS detection in above ground parts of the plant, that is, leaves and stems, based on spraying of 4-MUG was developed (Figure 1) (Martin *et al.*, in press).

Nondestructive Assay Protocol

1. Grow in soil or vermiculite.
2. Spray 1 ml 0.5–10 m*M* 4-MUG (predissolved in DMF) in Na-phosphate buffer, pH 7 (approximate amount per 5 seedlings), 0.05% Sapogenat T110 (Hoechst AG, Frankfurt).
3. Detection under UV light.
4. Incubate plants in dark at 37°C.
5. Evaluate after different periods of time.
6. Document on Polaroid with Wratten 2E filter.

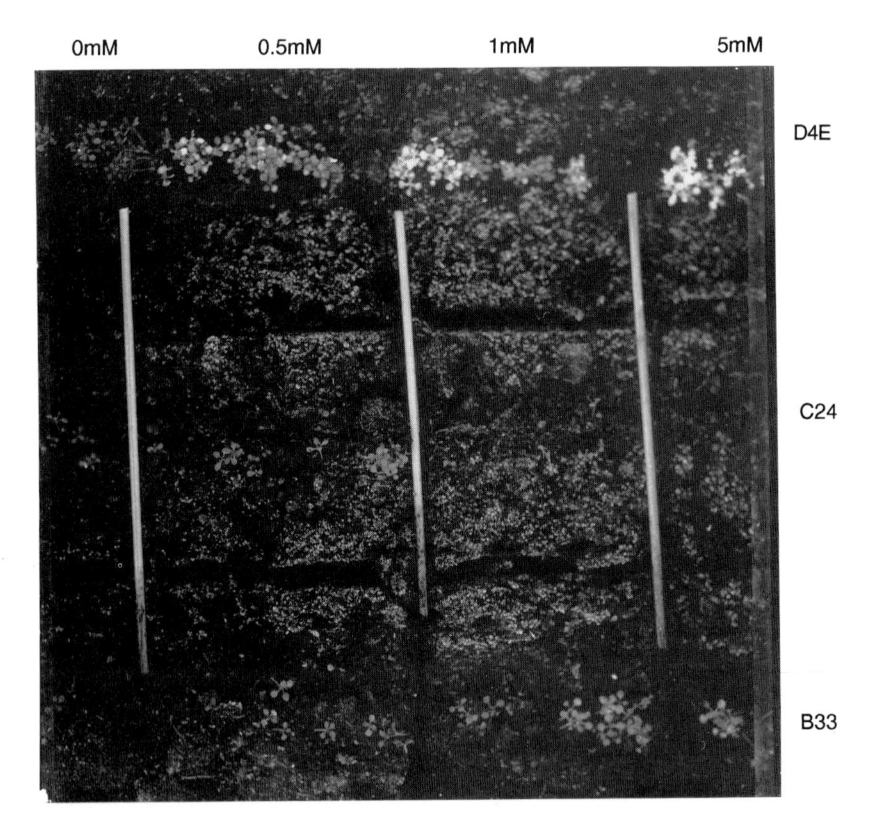

Fig. 1 Nondestructive MUG assay. Wild-type and transgenic (B33-GUS,D4E-GUS) *Arabidopsis* plants were grown for 2 weeks in soil. 4-MUG was sprayed and fluorescence was determined as described in text.

Comments

Depending on the plant condition, patchy staining patterns can be observed. Plants grown in tissue culture give a more uniform pattern. As the product 4-MU is diffusible in the plant, qualitative expression is difficult to analyze. Therefore analysis should be performed as early as possible. For all-or-nothing determinations plants can be incubated overnight.

Higher concentrations of Sapogenat lead to faster staining (at 0.5% within 3 h for D4E-GUS) but decrease the survival rates. Long irradiations quench the fluorescence, reduce survival, and probably induce mutations. UV light with an emission maximum at 365 nm should be less harmful compared to 305 nm and is closer to the absorption optimum of 4-MU. Resorufin-β-D-glucuronide (ReG) could be superior to 4-MUG because plant does not heavily absorb or fluoresce in 572/583 nm range.

GUS as a Tool to Study Development

For some purposes the toxicity of the product generated by GUS might be used to study development. By using promoters that can direct GUS to specific cell types in combination with the application of substrates that yield toxic products after cleavage by β-glucuronidases, the function and role of these cells in plant development can be studied. For this purpose young seedlings can be incubated overnight in liquid MS media (Murashige and Skoog, 1962) containing either X-gluc or other substrates, which yield toxic products after cleavage by GUS. The staining is lethal for the appropriate cells, and the effect of loss of these cells can be studied during the further development of the plant (R. Schmidt and W. B. Frommer, unpublished observation).

Is Vascular Tissue Preferentially Stained by GUS?

As in several studies the histochemical staining of transgenic plants was found to be vascular-associated, we have put together a list of promoter–GUS fusions that have been analyzed on the histochemical level (Table 2). We did not differentiate between xylem and phloem. Only in a few cases no staining of the vascular system could be found. Whether the observed data derive from an intrinsic function of the reporter/detection system or simply are coincidential is not possible to determine to date. Explanations for the preferential staining include kinetic arguments such as accumulation of the substrate or the product,

Table 2

Vascular Association of Expression Found for Different Promoter–Gus Fusions

Promoter–GUS fusion	Citation	Vascular-associated expression in organ/tissue	Expression level[a]		
			Major	Equal	No/low
Class I-patatin	Liu *et al.* (1990)	Young tubers	+	+	
Class II-patatin	Köster-Töpfer *et al.* (1989)	Tubers	+		
CaMV 35S	Jefferson (1987) Benfey *et al.* (1990a,b)	Roots and leaves	+	+	
Proteinase inh. II	Keil *et al.* (1990)	Tubers and leaves	+		
Sucrose synthase	Yang and Russell (1990)	Roots, leaves, stems, and flowers	+		
rbcs	Yang and Russell (1990)				+
Heat shock	Yang and Russell (1990)	Stems, leaves	+	+	
SAM synthetase	Pelemen *et al.* (1989)	Leaves, stems, and roots	+		
Zein	Schernthaner *et al.* (1988)				+
PAL genes	Liang *et al.* (1989) Bevan *et al.* (1989)	Flowers		+	
Chalcone synthase	Koes *et al.* (1990)	Leaves, stems, and roots	+		
Glycine-rich protein	Keller *et al.* (1989a)	Roots, stems, leaves, and flowers	+		
Hemoglobin	Bogusz *et al.* (1990)	Roots	+		
PRla	Ohshima *et al.* (1990)	Leaves	+		
Em	Marcotte *et al.* (1989)	Embryo, not analyzed in fully grown			
ST-LSl	Stockhaus *et al.* (1989)	Stem	+	+	
rolA, rolB, rolC	Schmülling *et al.* (1989)	Leaves, stems, and roots	+		
β-Phaseolin	Bustos *et al.* (1989)	Embryo	+		
atsla	Peleman *et al.* (1989)	Leaves and stems		+	
Glutamine synthetase	Forde *et al.* (1989)	Nodules	+		
Endogenous GUS	T.M., W.B.F. (unpublished)		+		

[a] The staining intensity relative to other cell types of the same organ is given as "major" and "equal."

local enhancement or inhibition of the enzyme, and the accessibility or permeability of the substrate in the tissue. To test the accessibility of the the substrate, CaMV 35S promoter/GUS fusions can be used as a positive control, though the expression from this promoter is not equal in all tissues.

Cell size and vacuolization could influence the result. For example, sieve-element companion cells are usually smaller compared to mesophyll cells in leaves or storage parenchyma cells in the potato tuber. A comparable expression level in the neighboring cells will lead to a stronger staining in the vascular system. Assuming that GUS is localized in the cytoplasm, small active cells with few small vacuoles, such as meristematic cells, will mimic strong expression when compared to large, highly vacuolated cells. The use of other targeting signals, for example, into the vacuole, should lead to the same results as for the cytoplasmic localization if the vacuolization is of no importance.

In order to confirm data obtained by GUS staining, independent methods such as *in situ* hybridization or localization using antibodies should be used, as in case of the vascular association of the cell wall glycine-rich protein (Keller *et al.,* 1989b).

Acknowledgments

We want to thank Erwin Heberle-Bors and W. Schäfer for sharing data prior to publication, Thomas Altmann for advice and support in spraying 4-MUG, and Peter Morris and Burkhard Schulz for critical reading of the manuscript.

References

Benfey P. N., Ren, L., and Chua, N. H. (1990a). Tissue-specific expression from CaMV 35S enhancer subdomains in early stages of plant development. *EMBO J.* 9:1677–1684.

Benfey, P. N., Ren, L., and Chua, N. H. (1990b). Combinatorial and synergistic properties of CaMV 35S enhancer subdomains. *EMBO J.* 9:1685–1690.

Bevan, M. (1984). Binary *Agrobacterium* vectors for plant transformation. *Nucleic Acids Res.* 12:8711–8721.

Bevan, M., Shufflebottom, D., Edwards, K., Jefferson, R., and Schuch, W. (1989). Tissue- and cell-specific activity of a phenylalanine ammonia-lyase promoter in transgenic plants. *EMBO J.* 8:1899–1906.

Bogusz, D., Llewellyn, D. J., Craig, S., Dennis, E. S., Appleby, C. A., and Peacock, W. J. (1990). Nonlegume hemoglobin genes retain organ-specific expression in heterologous transgenic plants. *Plant Cell* 2:633–641.

Bradford, M. M. (1976). A rapid and sensitive method for the quantitation of microgram quantities of protein utilizing the principle of protein-dye binding. *Anal. Biochem.* 72:248–254.

Bustos, M. M., Guiltinan, M. J., Jordano, J., Begum, D., Kalkan, F. A., and Hall, T. C. (1989). Regulation of β-glucuronidase expression in transgenic tobacco plants by an A/T-rich, cis-acting sequence found upstream of a French bean β-phaseolin gene. *Plant Cell* 1:839–853.

Damm, B., Halfter, U., Altmann, T., and Willmitzer, L. (1991). Transgenic *Arabidopsis*. In "Transgenic Plants," (eds. S. D. Kung and R. Wu), Butterworths, London.

Evans, P. T., and Malmberg, R. L. (1989). Do polyamines have roles in plant development? *Annu. Rev. Plant Physiol. Plant Mol. Biol.* 40:235–269.

Feldmann, K. A., Marks, M. D., Christianson, M. L., and Quatrano, R. S. (1989). A dwarf mutant of *Arabidopsis* generated by T-DNA insertion mutagenesis. *Science* 243:1351–1354.

Forde, B. G., Day, H. M., Turton, J. F., Wen-jun, S., Cullimore, J. V., and Oliver, J. E. (1989). Two Glutamine synthetase genes from *Phaseolus vulgaris* L. display contrasting developmental and spatial patterns of expression in transgenic *Lotus corniculatus* plants. *Plant Cell* 1:391–401.

Freeling, M. (1973). Simultaneous induction by anaerobiosis or 2,4-D of multiple enzymes specified by two unlinked genes: Differential Adh1-Adh2 expression in maize. *Mol. Gen. Genet.* 127:215–227.

Gould, J. H., and Smith, R. H. (1989). A non-destructive assay for GUS in the media of plant tissue cultures. *Plant Mol. Biol. Rep.* 7:209–216.

Hu, C. Y., Chee, P. P., Chesney, R. H., Zhou, J. H., Miller, P. D., and O'Brien, W. T. (1990). Intrinsic GUS-like activities in seed plants. *Plant Cell Rep.* 9:1–5.

Janssen, B. J., and Gardner, R. C. (1989). Localized transient expression of GUS in leaf discs following cocultivation with *Agrobacterium*. Plant Mol. Biol. 14:61–72.

Jefferson, R. A. (1987). Assaying chimeric genes in plants: The GUS gene fusion system. *Plant Mol. Biol. Rep.* 5:387–405.

Jefferson, R. A. (1989). The GUS reporter gene system. *Nature* 342:837–838.

Jefferson, R. A., Burgess, S. M., Hirsh, D. (1986). β-glucuronidase from *Escherichia coli* as a gene-fusion marker. *Proc. Natl. Acad. Sci. USA* 83:8447–8451.

Jefferson, R. A., Kavanagh, T. A., and Bevan, M. W. (1987). GUS fusions: β-Glucuronidase as a sensitive and versatile gene fusion marker, *EMBO J.* 6:3901–3908.

Jones, J. D. G., Garland, F. M., Maliga, P., and Dooner, H. K. (1989). Visual detection of transposition of the maize element Activator (Ac) in tobacco seedlings. *Science* 244:204–207.

Keil, M., Sanchez-Serrano, J. J., and Willmitzer, L. (1989). Both wound-inducible and tuber-specific expression are mediated by the promoter of a single member of the potato proteinase inhibitor II gene family. *EMBO J.* 8:1323–1330.

Keller, B., Schmid, J., and Lamb, C. J. (1989a). Vascular expression of a bean cell wall glycine-rich protein-β-glucuronidase gene fusion in transgenic tobacco. *EMBO J.* 8:1309–1314.

Keller, B., Templeton, M. D., and Lamb, C. J. (1989b). Specific localization of a plant cell wall glycine-rich protein in protoxylem cells of the vascular system. *Proc. Natl. Acad. Sci. USA* 86:1529–1533.

Koes, R. E., van Blokland, R., Quattrocchio, F., van Tunen, A. J., and Mol, J. N. M. (1990). Chalcone synthase promoters in petunia are active in pigmented and unpigmented cell types. *Plant Cell* 2:379–392.

Köster-Töpfer, M., Frommer, W., Rocha-Sosa, M., Rosahl, S., Schell, J., and Willmitzer, L. (1989). A class II patatin promoter is under developmental control in both transgenic potato and tobacco plants. *Mol. Gen. Genet.* 219:390–396.

Köster-Töpfer, M., Frommer, W. B., Rocha-Sosa, M., and Willmitzer, L. (1990). Presence of a transposon-like element in the promoter region of an inactive patatin gene in *Solanum tuberosum* L. *Plant Mol. Biol.* 14:239–247.

Kozak, M. (1989). The scanning model for translation: An update. *J. Cell Biol.* 108:229–241.

Kosugi, S., Ohashi, Y., Nakajima, K., and Arai, Y. (1990). An improved assay for β-glucuronidase in transformed cells: Methanol almost completely suppresses a putative endogenous β-glucuronidase activity. *Plant Sci.* 70:133–140.

Laemmli, U. K. (1970). Cleavage of structural proteins during the assembly of the head of bacteriophage T4. *Nature* 227:680–685.

Liang, X., Dron, M., Schmid, J., Dixon, R.A., and Lamb, C. J. (1989). Developmental and environmental regulation of the phenyalanine ammonia lyase β-glucuronidase gene fusion in transgenic tobacco. *Proc. Natl. Acad. Sci. USA* 86:9284–9288.

Liu, X. J., Prat, S., Willmitzer, L., and Frommer, W. (1990). Cis regulatory elements directing tuber specific and sucrose inducible expression of a class I patatin promoter. *Mol. Gen. Genet.* 223:401–406.

Liu, X. J., Rocha-Sosa, M., Rosahl, S., Willmitzer, L., and Frommer, W. (1991). A detailed study of regulation and evolution of the two classes of patatin genes. *Plant Mol. Biol.,* in press.

Logemann, J., Schell, J., and Willmitzer, L. (1987). Improved method for the isolation of RNA from plant tissues. *Anal. Biochem.* 163:16–20.

Lowry, O. H., Rosebrough, N. J., Farr, A. L., and Randall, R. J. (1951). Protein measurement with the Folin phenol reagent. *J. Biol. Chem.* 193: 265–275.

Marcotte, W. R., Jr., Russell, S. H., and Quatrano, R. S. (1989). Abscisic acid-responsive sequences from the Em gene of wheat. *Plant Cell* 1:969–976.

Martin, T., Schmidt, R., Altmann, T., Willmitzer, L., Frommer, W. Nondestructive assay systems for β-glucuronidase activity in higher plants. *Plant Mol. Biol. Rep.,* in press.

Murashige, T., and Skoog, F. (1962). A revised medium for rapid growth and bioassays with tobacco tissue cultures. *Physiol. Plant.* 15:473–497.

Ohshima, M., Itoh, H., Matsuoka, M., Murakami, T., and Ohashi, Y. (1990). Analysis of stress-induced or salicylic acid-induced expression of the pathogenesis-related 1a protein gene in transgenic tobacco. *Plant Cell* 2:95–106.

Olsson, O., Koncz, C., and Szalay, A. (1988). Use of the luxA gene of the bacterial luciferase operon as a reporter gene. *Mol. Gen. Genet.* 215:1–9.

Ow, D., Wood, K. V., DeLuca, L., DeWet, J., Helsinki, D., and Howell, S. H. (1986). Transient and stable expression of the firefly luciferase gene in plant cells and transgenic plants. *Science* 234:856–859.

Peleman, J., Boerjan, W., Engler, G., Seurinck, J., Botterman, J., Alliotte, T., Van Montagu, M., and Inzé, D. (1989). Strong cellular preference in the expression of a housekeeping gene of *Arabidopsis thaliana* encoding *S*-adenosylmethionine synthetase. *Plant Cell* 1:82–93.

Plegt, L., and Bino, R. J. (1989). β-glucuronidase activity during development of the male gametophyte from transgenic and non-transgenic plants. *Mol. Gen. Genet.* 210:321–327.

Raineri, D. M., Bottino, P., Gordon, M. P., and Nester, E. W. (1990). *Agrobacterium*-mediated transformation of rice (*Oryza sativa* L.). *Bio/Technology* 8:33–38.

Rocha-Sosa, M., Sonnewald, U., Frommer, W., Stratmann, M., Schell, J., and Willmitzer, L. (1989). Both developmental and metabolic signals activate the promoter of a class I patatin gene. *EMBO J.* 8:23–29.

Sambrook, J., Fritsch, and Maniatis, T. (1989). "Molecular Cloning: A Laboratory Manual," Cold Spring Harbor Laboratory Press, Cold Spring Harbor, N.Y.

Schäfer, W., Stahl, D., and Mönke, E. (1991). Identification of fungal genes involved in plant pathogenesis and host range. In "Advances in Plant Gene Research," Vol. 8, "Genes Involved in Plant Defence" (F. Meins and T. Boller, eds.), Springer Verlag Wien, New York.

Schernthaner, J. P., Matzke, M. A., and Matzke, A. J. M. (1988). Endosperm-specific activity of a zein gene promoter in transgenic tobacco plants. *EMBO J.* 7:1249–1283.

Schmitz, U. K., Hodge, T. P., Walker, E., and Lonsdale, D. M. (1990). Targeting proteins to plant mitochondria. In "Plant Gene Transfer 129," (C. Lamb and R. Beachy, eds.), pp. 237–248. Alan R. Liss, New York.

Schmülling, T., Schell, J., and Spena, A. (1989). Promoters of the rol A, B, and C genes of *Agrobacterium rhizogenes* are differentially regulated in transgenic plants. *Plant Cell* 1:665–670.

Stockhaus, J., Schell, J., and Willmitzer, L. (1989). Correlation of the expression of the photosynthetic gene ST-LS1 with the presence of chloroplasts. *EMBO J.* 8:2445–2451.

Tobin, E.M., and Silverthorne, J. (1985). Light regulation of gene expression in higher plants. *Annu. Rev. Plant Phys.* 36:569–593.

Töpfer, R., Pröls, M., Schell, J., and Steinbiss, H. H. (1988). Transient gene expression in tobacco protoplasts: II. Comparision of the reporter gene systems for CAT, NPTII, and GUS. *Plant Cell Rep.* 7:225–228.

Vancanneyt, G., Schmidt, R., O'Connor-Sanchez, A., Willmitzer, L., and Rocha-Sosa, M. (1990). Construction of an intron-containing marker gene: Splicing of the intron in transgenic plants and its use in monitoring early events in *Agrobacterium*-mediated plant transformation. *Mol. Gen. Genet.* 220:245–250.

Walden, R., and Schell, J. (1990). Techniques in plant molecular biology-progress and problems. *Eur. J. Biochem.* 192:563–576.

Willmitzer, L. (1988). The use of transgenic plants to study plant gene expression. *Trends Genet.* 4:13–18.

Yang, N. S., and Russell, D. (1990). Maize sucrose synthase-1 promoter directs phloem cell-specific expression of GUS gene in transgenic tobacco plants. *Proc. Natl. Acad. Sci. USA* 87:4144–4148.

PART 2 The GUS Assay

3 Quantitation of GUS Activity by Fluorometry

Sean R. Gallagher
Hoefer Scientific Instruments
San Francisco, California

This chapter gives a general overview of both fluorometric assays and the instrumentation required to quantitate β-glucuronidase activity (GUS) with fluorescence. A laboratory exercise covering the use of the most common fluorescent GUS assay is included. The exercise is intended for both student and basic research laboratories as it illustrates measurement of GUS activity and the determination of basic enzyme kinetic parameters of the GUS enzyme. A list of suppliers of fluorometers and reagents is also included in the appendix. Rendell (1987) is an excellent introduction to fluorometric assays and fluorescence instrumentation and should be consulted for further details.

The GUS Fluorometric Assay and Fluorescence Instrumentation

Although spectrophotometric substrates for GUS are available (see Naleway, Chapter 4, and Wilson *et al.*, Chapter 1), GUS activity in solution is usually measured with the fluorometric substrate 4-methylumbelliferyl-β-D-glucuronide (MUG). Fluorometry is preferred over spectrophotometry because of its greatly increased sensitivity and wide dynamic range. The assay is in addition highly reliable and simple to use. Occasionally, endogenous compounds will interfere with the

GUS Protocols: Using the GUS Gene as a Reporter of Gene Expression

assay, either by quenching or by producing a high background fluorescence (e.g., Martin *et al.*, Chapter 2). In these situations, fluorometric substrates with differing excitation and emission wavelengths are available [e.g., 4-trifluoromethylumbelliferyl β-D-glucuronic acid (4-TFMUG) or resorufin β-D-glucuronic acid; see Naleway, Chapter 4; Rao and Flynn, Chapter 6]. 4-TFMUG also allows continuous monitoring of GUS activity because, unlike MUG, it becomes fluorescent upon hydrolysis at the assay pH. In contrast, after hydrolysis of MUG by GUS, the reaction first must be terminated with a basic solution. This not only stops the enzyme reaction, but also causes the fluorescent moiety, 7-hydroxy-4-methylcoumarin (4-methylumbelliferone; MU), to become fully fluorescent. The properties of the above substrates are explained in detail by Naleway (Chapter 4).

Certain compounds give off light when excited by light of a shorter wavelength. When the excited state is very short (10^{-8} s) the light emitted as the compound returns to the ground state is referred to as fluorescence. Fluorescence is always at a longer wavelength than the excitation light (see Figure 2). For example, the GUS substrate MUG will not fluoresce until hydrolyzed by the enzyme. However, once the fluorescent moiety 4-methylumbelliferone (MU) is free in solution, it has a peak excitation of 365 nm (UV) and a peak emission of 455 nm (blue). In order to read the fluorescence of this GUS assay, the fluorometer must be capable of exciting and reading at or near these wavelengths.

Specialized instrumentation is required to measure fluorescence. In contrast to spectrophotometers, fluorometers select two wavelengths of light (Figure 1). The selection is made with either filters or emission monochromators. The excitation wavelength is analogous to absorption wavelength used in spectrophotometry, while the emission monochromator selects for the peak wavelength of fluorescence in the sample. Depending on the number of samples and level of automated data analysis needed, quantitation is possible using instruments ranging from the most basic and inexpensive filter fluorometer to highly sophisticated and very expensive scanning fluorometers. An important variation on this is the microtiter plate-reading fluorometers (Flow Laboratories; Dynatech Laboratories; Perkin-Elmer Corp.). Used by Rao and Flynn (Chapter 6), the microtiter plate format simplifies both automation and rapid processing of large numbers of samples. If the excitation and emission peaks are known for an assay, then instruments that select the wavelengths with filters in place of scanning monochromators are more than adequate and much less expensive. The filters can be interchangeable, or fixed at certain wavelengths. Filter fluorometers are

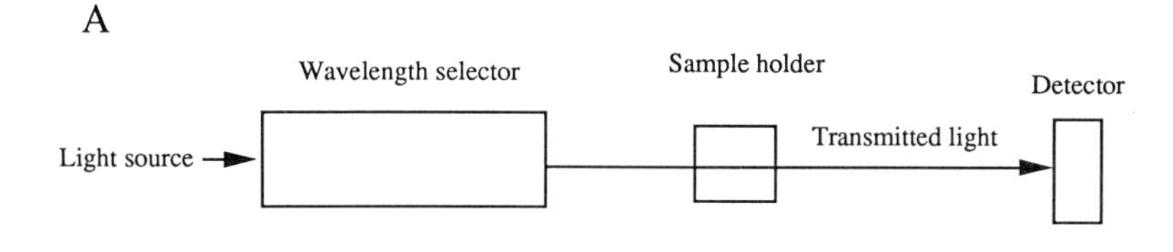

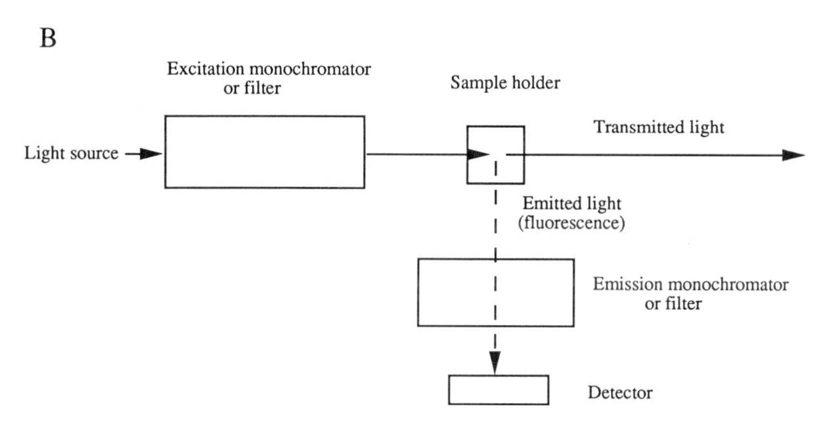

Fig. 1 Basic components of a spectrophotometer (A) and a fluorometer (B). See text for details.

very sensitive and are also are much simpler to use for routine GUS assays.

When initially characterizing a fluorescent molecule, two spectra are obtained. The excitation spectrum shows where the peak excitation wavelength is, and is performed by keeping the emission detector at or near the peak emission wavelength while scanning through the excitation wavelengths. The excitation spectra should be identical to the absorption spectra obtained with a spectrophotometer, although differences are sometimes caused by the instrumentation. The emission spectra is determined by keeping the excitation light fixed at the peak excitation wavelength and scanning the detector across the emission wavelengths. Note that the two spectra are near mirror images of each other (Figure 2; see also Naleway, Chapter 4). The magnitude of the distance between the excitation and emission peaks is referred to as the

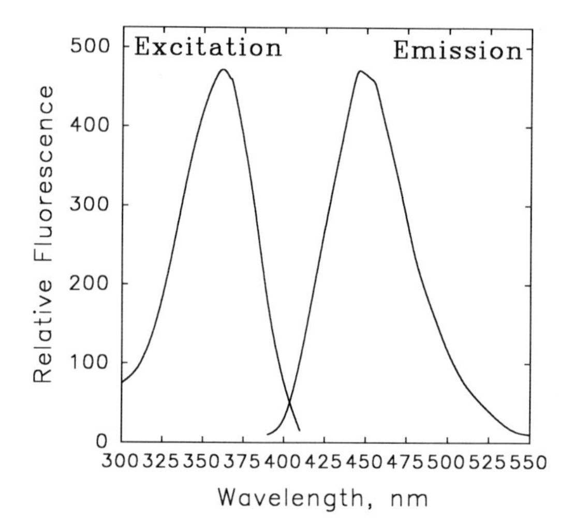

Fig. 2 Excitation and emission spectra of a solution of 10 n*M* 4-methylumbelliferone (MU) in 2.0 ml carbonate stop buffer (0.2 *M* Na_2CO_3).

Stokes' shift and indicates how well resolved the emission light will be from the excitation light. For commercially available fluorochromes this information is usually available. Scans of fluorochromes used in GUS assays are presented by Naleway (Chapter 4).

Typically, fluorometers have a right-angle (90°) geometry in which the excitation light enters one side of the cuvette and fluorescence is measured by reading the adjacent side (Fig. 1). Thus, unlike a spectrophotometer, the excitation light does not directly impinge on the emission detector. Other geometries are also possible for specialized applications. This includes frontal illumination in which the excitation light enters and the emission light is detected through the same face of the cuvette. In these cases the sample is so highly turbid or absorptive that the excitation light does not penetrate far into the solution and the fluorescence level does not represent the true concentration of the fluorochrome if the standard 90° geometry is used. Fluorescein mono-β-D-glucuronide requires frontal illumination (Naleway, Chapter 4).

Light sources also differ between fluorometers and spectrophotometers. For absorption measurements, relatively low-power deuterium lamps are common. High-intensity light can actually be problematic in spectrophotometers due to photodecomposition and fluorescence of the sample. By comparison, because the amount of fluorescence is directly proportional to the level of the excitation light, fluorometers achieve extremely high sensitivity and a wide dynamic range through use of high-intensity light sources. Dynamic range is the highest concentration divided by the lowest concentration that can be measured under standard conditions. Spectrophotometers have a dynamic range of around 100, while fluorometers can achieve dynamic range of 10,000 or more. Fluorometers are, however, limited to relatively low concentrations of fluorochrome. For a linear calibration graph, absorbance of the highest concentration of the fluorochrome should not exceed 0.02. Instruments capable of selecting both UV and visible excitation wavelengths generally use a xenon arc lamp for a continuous light spectrum. Mercury vapor lamps are an inexpensive alternative and are found in filter-based instruments dedicated to more limited ranges of wavelengths.

Even though fluorescence is emitted in all directions, the light is usually measured at a 90° angle from the excitation beam as described above. Because of the requirement to have two adjacent optically clear sides, sample cuvettes are specifically designed for use in fluorometers. Spectrophotometer cuvettes are not recommended. Semimicro spectrophotometer cuvettes, for example, have two opposite frosted sides and cannot be used in a fluorometer. Typically, fluorometer cuvettes have all four sides and the bottom polished and optically clear. It is important to keep these surfaces clean and free from scratches

The GUS Assay: A Laboratory Exercise

The following procedures describe the fluorometric assay based on the hydrolysis of the MUG substrate by GUS, with the reaction

4-methylumbelliferyl β-D-glucuronide (MUG) → glucuronic acid + 7-hydroxy-4-methylcoumarin (MU)

The 7-hydroxy-4-methylcoumarin (4-methylumbelliferone; MU) is maximally fluorescent when the hydroxyl is ionized, and as a consequence the assays are stopped with sodium carbonate.

The emission and excitation spectra in Fig. 2 were generated with an RF5000 (Shimadzu) scanning spectrofluorometer. This rest of the exercise was performed with the TKO 100 (Hoefer Scientific Instruments), a filter fluorometer with the wavelengths fixed at 365 and 460 nm for excitation and emission, respectively. These wavelengths also permit quantitation of DNA using the DNA-binding fluorochrome H33258 (Gallagher, 1989). The assay described below can also be used with any fluorometer that has these excitation and emission wavelengths.

The K_m for the MUG substrate under the conditions listed in this bulletin is 0.57 m*M*, with a minimum detectable GUS activity using the TKO 100 of 1 n*M* MUG hydrolyzed in a final quenched assay volume of 2.0 ml. This is equivalent to 10 counts on the TKO 100 display or 2 pmol MU.

Materials

The TKO 100 (Hoefer, TKO 100-115V) fluorometer comes complete with square glass fluorometry cuvette (Hoefer, TKO105). Disposable plastic cuvettes specifically made for fluorometric measurements can be obtained from Sarstedt (67.755). The 0–2 ml RePipet Jr (Hoefer, TKO150) is useful for dispensing stop buffer at the end of the reaction.

The GUS gene is available in a variety of configurations from Clontech Laboratories. Materials for detection and analysis of the GUS gene are available from a number of suppliers including Clontech and Molecular Probes. See the appendix for a complete list of chemical and equipment suppliers.

The enzyme is β-glucuronidase (GUS, from *E. coli*, Clontech 2251-1; Boehringer Mannheim 127 680; Sigma Chemical Co. G 7896).

The substrate is 4-methylumbelliferyl-β-D-glucuronide (MUG, Clontech 8082; Molecular Probes M-1490; Boehringer Mannheim 270 954; Sigma Chemical Co. M 9130), with an anhydrous molecular weight of 352.3.

The calibration standard is 7-hydroxy-4-methylcoumarin, sodium salt (MU; 4-methylumbelliferone; β-methylumbelliferone; Clontech 8084; Molecular Probes H-189; Sigma Chemical Co. M 1508), with a molecular weight of 198.2.

The general reagents needed for the extraction and stop buffers are available from several companies, including Hoefer Scientific Instruments. The reagents used in this exercise were from BDH (Hoefer Scientific Instruments) (see appendix): Na_2HPO_4, (MW=141.96), NaH_2PO_4 (MW=119.98), Na_2CO_3 (MW=105.99), β-mercaptoethanol, $Na_2EDTA \cdot 2H_2O$ (MW=372.24), sodium lauryl sarcosine (30% solution), and Triton X-100.

Methods

Solutions (Jefferson, 1987; Jefferson and Wilson, 1991)

Carbonate stop buffer: 0.2 *M* Na_2CO_3. For 1 liter, mix 21.2 g of Na_2CO_3 in a final volume of 1 liter distilled water.

Concentrated MU calibration stock solution: 1 μM 7-hydroxy-4-methylcoumarin (MU) in distilled water. To prepare MU standard, first, mix 19.8 mg 7-hydroxy-4-methylcoumarin, sodium salt, into 100 ml distilled water to produce a 1 m*M* MU solution. Dilute 10 μl of the 1 m*M* MU solution into 10 ml distilled water to make the 1 μM MU stock. Store at 0–5°C protected from light.

MU calibration standard: 50 n*M* MU. To prepare the 50 n*M* MU calibration standard, dilute 100 μl of the 1 μM concentrated MU calibration stock solution into 1.9 ml carbonate stop buffer. This solution should be made fresh just before use.

GUS extraction buffer: 50 m*M* $NaHPO_4$ (pH 7.0), 10 m*M* β-mercaptoethanol, 10 m*M* Na_2EDTA, 0.1% sodium lauryl sarcosine, 0.1% Triton X-100. For 100 ml GUS extraction buffer:

1 *M* $NaPO_4$, pH 7.0: 5 ml
β-Mercaptoethanol: 0.07 ml
0.5 *M* Na_2EDTA, pH 8.0: 2 ml
30% Sarcosyl: 0.33 ml
10% Triton X-100: 1 ml
Distilled water: 91.6 ml

GUS assay buffer: 2 m*M* MUG in extraction buffer. To prepare 25 ml assay solution, mix 4-methylumbelliferyl β-D-glucuronide (22 mg) and extraction buffer (25 ml).

Please note that the water content of the MUG preparations will vary. For greatest accuracy, the calculation of solution molarity should take this into account.

Calibration

The TKO 100 is adjusted with the scale knob so that 50 n*M* MU in stop buffer equals 500 counts on display (i.e., 10 display counts/n*M*). Scanning spectrofluorometers such as the Shimadzu RF-5000 and Perkin-Elmer LS 50 offer a variety of features for calibration. This includes data storage and automated creation and analysis of the calibration curve. As with the fluorometer used in this exercise, the curve generated with MU can be corrected for the background simply by subtracting the blank from the readings. For use with other fluorometers, manufacturers' instructions for instrument calibration and correction for the blank should be consulted.

1. Turn the TKO 100 on for at least 15 min before use.
2. Fill a clean glass cuvette with 1.9 ml of carbonate stop buffer (without MU). Place the cuvette into the sample chamber and close the lid. Note that the "G" mark on the top side of the cuvette should face forward. This insures that optical faces of the cuvette are in the same orientation with each reading.
3. With the scale knob set fully clockwise, bring the display to zero by adjusting the "Zero" knob on the TKO 100. This step corrects for the blank. Alternatively, note the blank value and subtract from the sample reading (e.g., Kyle *et al.*, Chapter 14).
4. Add 100 μl of the 1 μM MU standard to the carbonate blank, and mix by inversion to give the 50 nM MU standard solution. Place cuvette into sample chamber and close lid.
5. Adjust the scale knob so the TKO 100 reads 500. Thus a 50 nM solution of MU will read 500 on the TKO 100 display; that is, 10 display counts/nM MU.

A typical standard curve is shown in Figure 3. MU will photodegrade when exposed to intense excitation light, causing a downward drift in the reading. This is normally not a problem for mercury lamp/filter-based instruments like the TKO 100. However, with high-intensity xenon arc lamp sources this can be quite dramatic. Test for stability of the reading under your conditions by observing the reading over a period of 1–2 min. If downward drift is a problem, keep sample exposure to the excitation light to a minimum. Some fluorometers such as the Shimadzu RF5000 can automatically open and close a shutter so that the sample is only exposed to the excitation light during the reading.

Time Course Assay

The linearity of the assay with time is critical and should be verified under your specific conditions. The assay listed below is for a commercial suspension of GUS diluted into extraction buffer.

1. Take 10 μl of the commercial enzyme stock and dilute into 1 ml extraction buffer for a 1/100 dilution. Then take 10 μl of the first 1/100 dilution and add to 10 ml of extraction buffer to achieve the

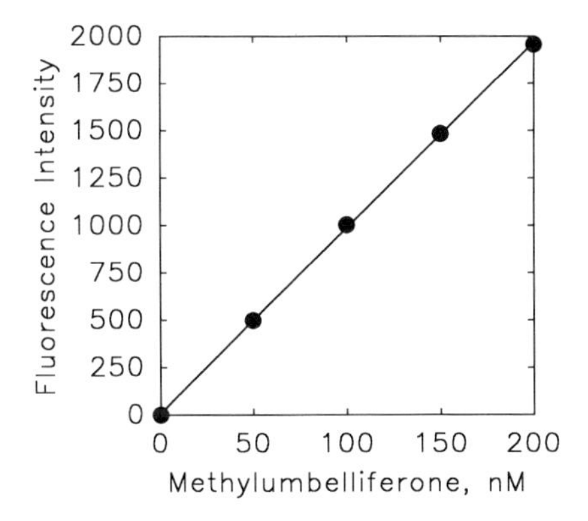

Fig. 3 Standard curve showing the linear rise in fluorescence with increasing concentrations of MU in carbonate stop buffer.

1/100,000 dilution. Keep enzyme stocks on ice. Use this final dilution for subsequent assays.

2. In two test tubes on ice, add:
   Assay solution: 250 μl
   Extraction buffer: 200 μl
   GUS enzyme (diluted): 50 μl
   This gives a final concentration of 1 mM MUG.
3. While still on ice, remove duplicate 50-μl aliquots for the reagent blank. These should be quenched immediately in 1.95 ml carbonate stop buffer.
4. Take the test tubes containing the assay solution from ice and start the assay by placing them in a 37°C water bath, staggered at 30-s intervals.
5. For each time point, remove 50-μl aliquots from both test tubes at 30-s intervals. Immediately quench the reaction by placing into 1.95 ml stop buffer.
6. Read samples in a calibrated fluorometer as described in section above. Figure 4 illustrates a typical time course.

Enzyme Kinetics

The Lineweaver-Burk enzyme kinetic plot (Segel, 1976) described in this section is routinely used by researchers for preliminary characterization of enzyme kinetics. Two of these parameters, K_m and V_{max}, are determined in the following experiment. The K_m is the concentration of

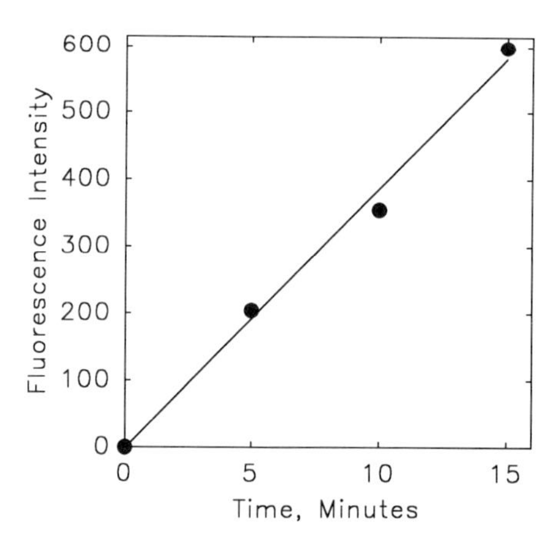

Fig. 4 Time course of the GUS assay. Each time point represents duplicate 50 μl aliquots quenched in 1.95 ml of carbonate stop buffer.

substrate at which the enzyme is at half maximal velocity and is a useful number for several reasons. Physiologically, the K_m indicates whether or not the enzyme efficiently uses a natural or artificial substrate. The concentration of the enzyme's substrate within a cell should be approximately equivalent to the K_m; extremes of [S] provide very little control of enzyme activity for the cell. At $[S] << K_m$, the enzyme activity is at a fraction of its potential maximum and is very sensitive to increases in [S], while at $[S] >> K_m$, changes in [S] have very little effect on activity (see Figure 5).

From a practical point of view, the K_m also indicates the appropriate concentration range of the substrate to use in an *in vitro* assay. As the enzyme hydrolyzes the substrate, the [S] drops during the assay, and this decrease in concentration can have a marked effect on the linearity of the assay. For example, if the [S] is too low ($[S] << K_m$) the enzyme activity will drop during the assay due to the decreasing [S] and cause the assay to become nonlinear with time. Starting at $[S] >> K_m$ keeps the enzyme near its maximal activity (V_{max}) throughout the length of the assay, even though the [S] is dropping. This is illustrated in Figure 5. Note that the activity is very sensitive to changes in [S] when [S] is less than the K_m of 0.57 m*M*. In contrast, when the [S] is larger than the K_m, increasing the [S] has less and less effect on activity because the enzyme is near its maximal velocity or V_{max}. To insure the enzyme activity is kept maximal during the length of the assay, the [S] in an assay is approximately two to three times the K_m. In this way, even moderate decreases in the [S] will not appreciably change enzyme activity.

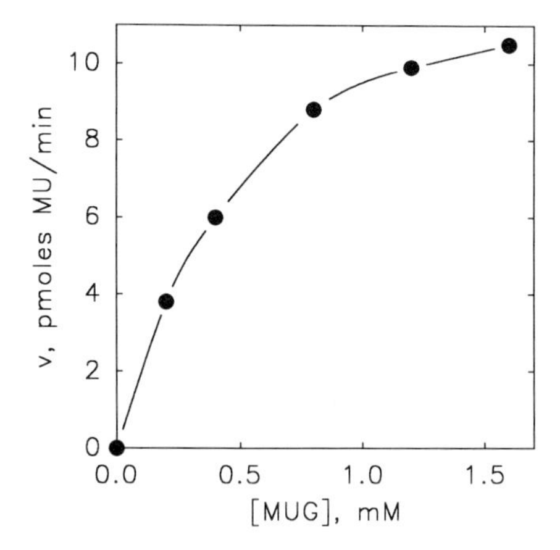

Fig. 5 GUS activity with increasing concentration of MUG substrate. Duplicate assays were run for each concentration. Assays minus enzyme were run in parallel at each concentration and used to correct for nonenzymatic hydrolysis of MUG.

1. Calibrate fluorometer as described above. Set TKO 100 to read 10 display units/n*M* MU.
2. In 20 test tubes—four at each concentration—prepare the 100-μl assay solutions according to Table 1.
3. Place the test tubes in a 38°C water bath. For each concentration, add 10 μl of extraction buffer without enzyme to two of the test tubes. These will serve as the blanks and correct for any nonenzymatic hydrolysis of MUG. Start the assay by adding 10 μl of the diluted GUS enzyme mixture (see Time Course Assay section) to the two remaining test tubes. Each assay, including the blanks, should be started a suitably timed interval such as 15 s. Alternatively, prepare the assay solutions as described above but

Table 1

Solutions for Determination of GUS K_m and V_{max}

[MUG] (Final)	Substrate Mix (2 m*M* MUG)	Extraction Buffer	Enzyme or Blank Solution
1.6 mM	80 μl	10 μl	10 μl
1.2	60	30	10
0.8	40	50	10
0.4	20	70	10
0.2	10	80	10

keep the test tubes on ice while adding the blank and enzyme solutions. With the assays kept at 0–4°C, the timing of the addition of the reagents is not critical. Start the assay by transferring the test tubes at 15-s intervals to the 38°C water bath.

4. After 10 min, quench the assays at 15-s intervals by adding 1.9 ml of the carbonate stop buffer.
5. Read each sample, average the values, and subtract the reagent blank.

Convert the readings to nanomoles MUG hydrolyzed by recalling that the readout is in 10 units per nanomolar MU in 2.0 ml stop buffer. Thus, a readout of 190 is equal to 19 n*M* MU in a final volume of 2.0 ml. To determine the total picomoles MU present in the 2.0 ml of stop solution, simply multiply 19×10^{-9} mol/liter $\times$ 0.002 liter to get 38×10^{-12} or 38 pmol. Divide by the assay time for a measure of enzyme activity in pmol MU/min. Data used in Figures 5 and 6 are illustrated in Table 2.

Lastly, convert the data to a more suitable format in order to calculate enzyme kinetics. A typical Lineweaver-Burk plot is shown in Figure 6. The plot is referred to as a double reciprocal because the data is plotted as 1/enzyme activity and 1/substrate concentration for the *y* and *x* axis, respectively (Segal, 1976). Linear regression is performed to calculate

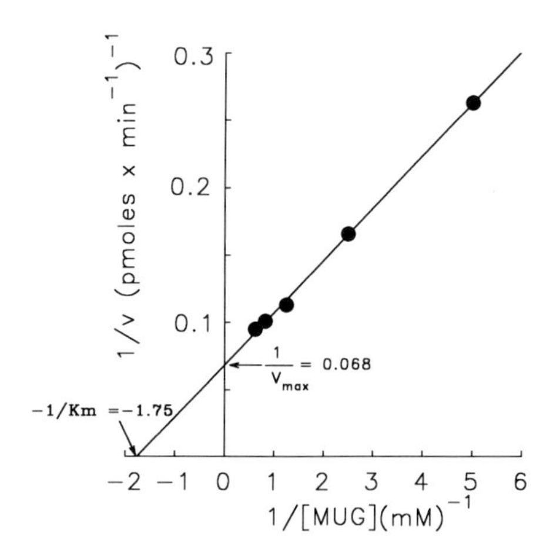

Fig. 6 Lineweaver-Burk double-reciprocal ($1/v$ vs 1/[S]) plot of GUS activity using the substrate MUG. Data from Fig. 5 was used in the plot. The km and V_{max} were estimated to be 0.57 m*M* and 14.7 pmol $\times$ min^{-1}, respectively.

Table 2

Double Reciprocal Values for Lineweaver-Burk Plot of GUS Activity

[MUG] mM	pmoles MU/min (10 min assay)	1/[MUG]	1/(pmoles MU/min)
0.2	3.8	5	0.263
0.4	6.04	2.5	0.166
0.8	8.81	1.25	0.114
1.2	9.92	0.83	0.101
1.6	10.54	0.625	0.095

V_{max} and K_m. The y-intercept value is $1/V_{max}$ and the X-intercept value is $1/K_m$.

This illustrates another important point about fluorometric substrates. So little enzymatic substrate hydrolysis is required to produce a large fluorescence signal that even low concentrations of MUG can be analyzed without markedly changing the concentration of MUG during the length of the assay. For example, only 0.2% of the total MUG present in the 0.2 m*M* assay solution was hydrolyzed in Figure 5, permitting the researcher to work well below the K_m of the enzyme without markedly affecting linearity.

References

Gallagher, S. R. 1989. Spectrophotometric and fluorometric quantitation of DNA and RNA in solution. In "Current Protocols in Molecular Biology" (F. A. Ausubel, R. Brent, R. E. Kingston, D. D. Moore, J. G. Seidman, J. A. Smith, and K. Struhl, eds.), pp. A.3.9–A.3.15. Greene Publishing and Wiley Interscience, New York.

Jefferson, R. A. 1987. Assaying chimeric genes in plants: The GUS gene fusion system. *Plant Mol. Biol. Rep.* 5:387–405.

Jefferson, R. A., and Wilson, K. J. (1991). The GUS gene fusion system. *In* "Plant Molecular Biology Manual" (S. B. Gelvin, R. A. Schilperoort, and D. P. S. Verma, eds.), pp. B14/1–B14/33. Kluwer, Boston.

Rendell, D. 1987. "Fluorescence and Phosphorescence." John Wiley & Sons, Chichester.

Segal, I. H. 1976. "Biochemical Calculations." John Wiley & Sons, New York.

4 Histochemical, Spectrophotometric, and Fluorometric GUS Substrates

John J. Naleway
Molecular Probes, Inc.
Eugene, Oregon

Use of the *Escherichia coli* β-glucuronidase (GUS) system as a reporter gene in plant molecular biology has helped spawn a new era of analysis and interpretation of the factors mediating the regulation of gene expression. Coinciding with the use of the GUS marker gene system has been the development of a series of GUS substrates and analytical methods for use in detecting enzyme expression levels.

Coexpression of β-glucuronidase (GUS) activity in recombinant plant cells has become a widely used technique for demonstration of exogenous gene fusions (Jefferson, 1987; Jefferson *et al.*, 1987). β-Glucuronidase vectors have been constructed for use with both *Agrobacterium*- and microprojectile-mediated DNA transfer methods (McCabe *et al.*, 1988; Jefferson *et al.*, 1987). Expression of β-glucuronidase gene fusions can be measured by fluorometric, spectrophotometric, or histochemical assay. GUS is quite stable and is active over a wide pH range. However, assays of the *E. coli* enzyme are generally performed between pH 7–8 to avoid the acid pH GUS or GUS-like activity present in many plant and animal tissues (see Martin *et al.*, Chapter 2; Stomp, Chapter 7; and Kyle *et al.*, Chapter 14). Most importantly, since the GUS gene is normally absent in plant tissues, there are typically no detectible background levels of β-glucuronidase activity in most higher plant cells.

Several fluorescent, spectrophotometric, or chromogenic GUS substrates are available commercially for assaying chimeric genes in plants using the GUS gene fusion system. This chapter will discuss these

GUS Protocols: Using the GUS Gene as a Reporter of Gene Expression

substrates with an emphasis on the methods used to maximize sensitivity of detection of GUS activity in plant tissues using fluorescent GUS assays.

The β-Glucuronidase Enzyme System

The β-glucuronidase enzyme from *E. coli* (EC 3.2.1.31) has been well documented to provide several desirable characteristics as a marker gene in transformed plants. First, the GUS gene from the *E. coli uidA* locus is quite stable to cloning techniques. It is stable to many detergents and a wide variety of ionic strength conditions. It has been successfully cloned into a great number of plant species via a variety of available plasmids (Jefferson, 1987; see also Farrell and Beachy, Chapter 9; Finnegan, Chapter 11; and Osbourn and Wilson, Chapter 10). The enzyme is capable of tolerating large amino-terminal additions for translational fusions. It has a monomeric molecular weight of approximately 68,000 but exists *in vivo* as a tetrameric species. It has a wide specificity range for β-conjugated glucuronides but will not cleave other glycosides, such as α- or β-glucoside substrates types (Tomasic and Reglevic, 1973).

Although β-glucuronidase is ubiquitous in animal tissue (see, however, Kyle *et al.*, Chapter 14, and Gallie *et al.*, Chapter 13), being lysosomal in location, it is not usually present in higher plant species. There are, however, exceptions, and in addition to the GUS-like background activity found in many plant tissues by Hu *et al.* (1990), activity has been reported in pollen grains (Sood, 1980) and in rye where the β-glucuronidase activity is highly substrate specific (Schulz, 1987).

The Types of Substrates

Various β-glucuronic acid substrates are available for detection of GUS expression *in vivo* and *in vitro*. All of these substrates contain the sugar D-glucopyranosiduronic acid (see Figure 1) attached by glycosidic linkage to a hydroxyl group (usually a phenolic hydroxyl) of a chromogenic, fluorogenic, or other detectible molecule.

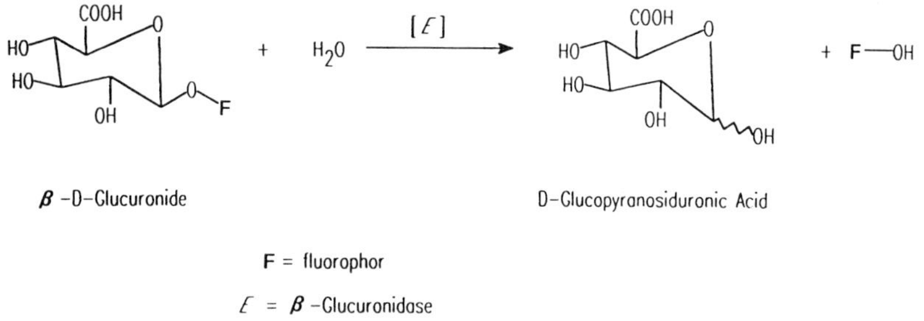

β-D-Glucuronide + H_2O →[E] D-Glucopyranosiduronic Acid + F—OH

F = fluorophor

E = β-Glucuronidase

Fig. 1 *Escherichia coli* β-glucuronidase specifically hydrolyzes β-linked D-glucuronides to D-glucuronic acid and aglycone.

5-Bromo-4-Chloro-3-Indolyl β-D-Glucuronic Acid (X-GlcU)

5-Bromo-4-chloro-3-indolyl β-D-glucuronide (X-GlcU) is generally recommended for histochemical localization of GUS activity and is commercially available from several companies (U.S. Biochemical, Cleveland, Ohio, product 12388; Molecular Probes, Eugene, Ore., catalog number B-1691; Sigma, St. Louis, Mo., catalog number B-0522; Clontech, Palo Alto, Calif.). Although this substrate provides only a colorimetric method of assaying GUS activity, precipitation of the product in aqueous solutions provides a convenient way of localizing enzyme activity *in vivo*. This colorless substrate produces a blue indigo dye precipitate at the site of enzymatic cleavage. Color formation requires three separate reactions. After enzymatic turnover, the released indoxyl derivative dimerizes and is then oxidized to the final indigo dye.

The staining protocol is well documented and easily performed (Pearson *et al.*, 1961). Improvements to these protocols that involve added oxidative reagents (Lojda, 1970) or fixing agents have improved the quality of the histochemical localization. See Martin *et al.* (Chapter 2), Stomp *et al.* (Chapter 7), and Craig (Chapter 8) for details or histochemical localization.

The assay procedure usually consists of soaking the tissue in the substrate solution for several minutes at room temperature and watching for the appearance of a blue color. The substrate is dissolved at a concentration of 2–10 m*M* in a suitable buffer (0.1 *M* phosphate buffer, pH 7.0) containing ethylenediamine tetraacetic acid (EDTA) (10 m*M*), and 0.5 m*M* each of potassium ferrocyanide and potassium ferricyanide. Care should be exhibited when preparing stock solutions since β-glucuronidase is inhibited by certain divalent cations (Cu^{2+}, Zn^{2+}, etc.) (Jefferson *et al.*, 1987). Stock solutions of X-GlcU can be prepared in advance and stored frozen (−20°C) until needed.

For tissue samples, enough staining solution is added to cover the tissue samples. Triton X-100 may be required to sufficiently wet some tissues. Incubation at 37°C overnight may be required to completely develop color. Tissue samples should be covered with plastic film to prevent drying. After the staining procedure is complete, the tissue is soaked in 95% ethanol to remove excess substrate. If sections of the stained tissue are required, the tissue is frozen at this point and then embedded. See Stomp (Chapter 7), Craig (Chapter 8), and Martin *et al.* (Chapter 2) for more details.

The use of histochemical localization to analyze gene expression is not without potential problems. In particular, accurate interpretation is dependent on such variables as the threshold of detection of the blue stain, cell size, and cell metabolic activity. In general, however, localization of GUS activity in plants using X-GlcU typically correlates well with results obtained using other methods (see below and Martin *et al.*, Chapter 2).

A distinct advantage to the use of X-GlcU is the ability to perform analysis on whole tissue samples or whole cells.

p-Nitrophenyl β-D-Glucuronide (pNPG)

para-Nitrophenyl glucuronide (Sigma, St. Louis, Mo., catalog number N-1627) is a standard spectrophotometric substrate for detection of β-glucuronidase activity *in vitro*. Detection of activity is a result of a shift in the absorption maximum of the phenol on cleavage of the glycosidic bond and release. The liberated 4-nitrophenol is measured spectrophotometrically at 402–410 nm, and absorbance intensity at these wavelengths relates directly to the specific activity.

Although *p*-nitrophenyl glucuronide is not a fluorogenic substrate, it is often used as a standard for enzymatic assays because of its low K_m value. However, because fluorescent detection techniques are typically several orders of magnitude more sensitive than spectrophotometric analyses, both the soluble nature and the lower sensitivity of this substrate make it a less widely used reagent for GUS analysis than X-GlcU or the fluorescent substrates listed below.

4-Methylumbelliferyl β-D-Glucuronic Acid (4-MUG)

The most widely used fluorogenic substrate for detection of β-glucuronidase activity *in vitro* is 4-methylumbelliferyl β-D-glucuronide

(4-MUG) (Fluka, Ronkonkoma, N.Y., catalog number 69602; Koch-Light, Ltd., Suffolk, U.K.; Molecular Probes, Eugene, Ore., catalog number M-1490; Sigma, St. Louis, Mo., catalog number M-9130; Clontech, Palo Alto, Calif.). Upon hydrolysis by GUS, however, the fluorochrome 4-methylumbelliferone (7-hydroxy-4-methyl coumarin) is produced along with sugar glucuronic acid. Using excitation at 363 nm and measuring emission at 447 nm, background fluorescence from the substrate is negligible (see Figure 2). It is important to note, however, that the substrate does exhibit a slight intrinsic fluorescence (em 375 nm) if excited at 316 nm. The excitation source and/or appropriate filter sets should be chosen, therefore, with this in mind. An example would be a DAPI/Hoechst filter set with excitation bandwidth at 365 ± 30 nm and emission at wavelengths greater than 420 nm. In general, measurements of GUS activity using this fluorescent substrate are two to three orders of magnitude more sensitive than X-GlcU or pNPG.

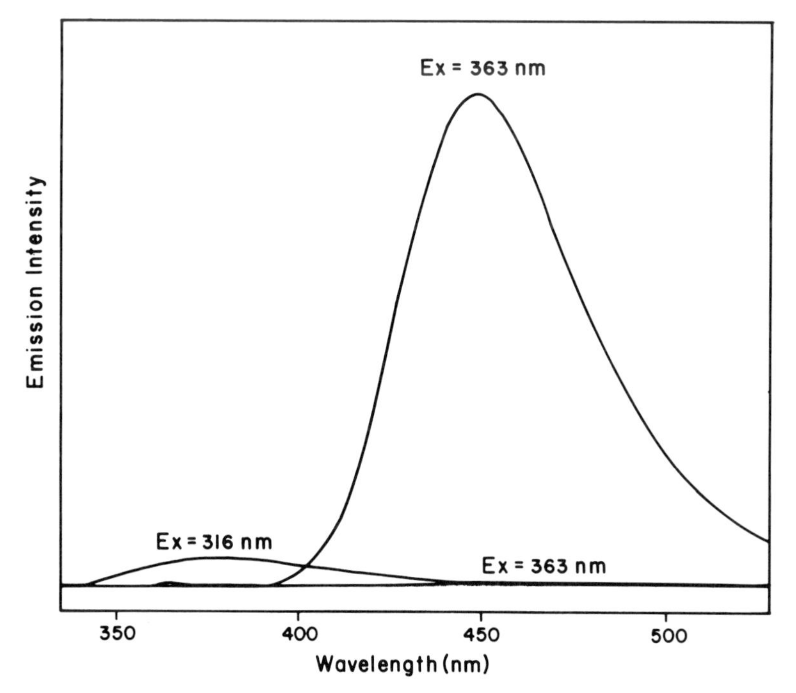

Fig. 2 Emission spectra for the substrate 4-methylumbelliferyl β-D-glucuronide (4-MUG) and its enzymatic hydrolysis product 4-methylumbelliferone (4-MU). The rate of enzymatic hydrolysis can be measured by the increase in fluorescence emission measured at 447 nm.

The product is itself a pH-sensitive fluorescent probe (pK_a 8.2) that exhibits maximal fluorescence at pH values above its pK_a (Chen, 1968; Gerson, 1982). At physiological pH values, the fluorescence is relatively low. This is due to the equilibrium between the phenolic form of the product and its phenoxide form at high pH values (see Figures 3 and 4).

It is the phenoxide form of the dye that gives maximal fluorescence at 447 nm. The phenolic form of the dye actually has different fluorescence properties (ex 323, em 386 nm). Treatment of samples with 0.2 *M* sodium carbonate buffer (pH 9.5) after incubation both quenches the enzyme reaction and raises the pH above the pK_a of 4-methylumbelliferone (4-MU) to produce the maximum amount of fluorescence.

The fluorogenic assay can be run either with homogenized crude tissue samples or on tissue or culture extracts. Lysis conditions for such samples have been described (Jefferson, 1987) in which an extraction buffer [50 m*M* sodium phosphate, pH 7, containing 10 m*M* dithiothreitol (DTT), 1 m*M* disodium EDTA, 0.1% sodium lauryl sarcosine, and 0.1% Triton X-100] is used. Removal of low-molecular-weight endogenous fluorescent compounds can be accomplished, if necessary, by treating these extracts with Polycar (insoluble polyvinyl pyrollidone) followed by centrifugation and a Sephadex G-25 column. These extracts should be stored frozen (−20°C) prior to use.

Typically, the assay is performed by incubating samples of extract in 1 m*M* 4-MUG solution at 37°C. At regular time intervals, aliquots are removed and added to the stop buffer (0.2 *M* sodium carbonate, pH 9.5).

Since the assay is based on measurements of relative fluorescence intensity, standard solutions of 4-methylumbelliferone (4-MU) should be used to calibrate the data during each run. These should be prepared in 0.2 *M* sodium carbonate solution (stop buffer). Using these standard solutions, a calibration curve will directly produce relative fluorescence data for samples in units of moles 4-MU. The relative fluorescence is read over a time course (5, 10, 15 min., etc.) and converted into nanomoles 4-MU produced per minute per milligram sample (see Gallagher,

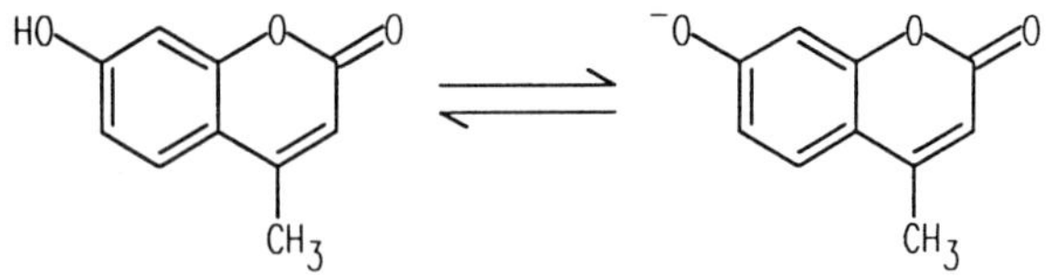

Fig. 3 The equilibrium between phenol and phenoxide forms of 4-methylumbelliferone (4-MU).

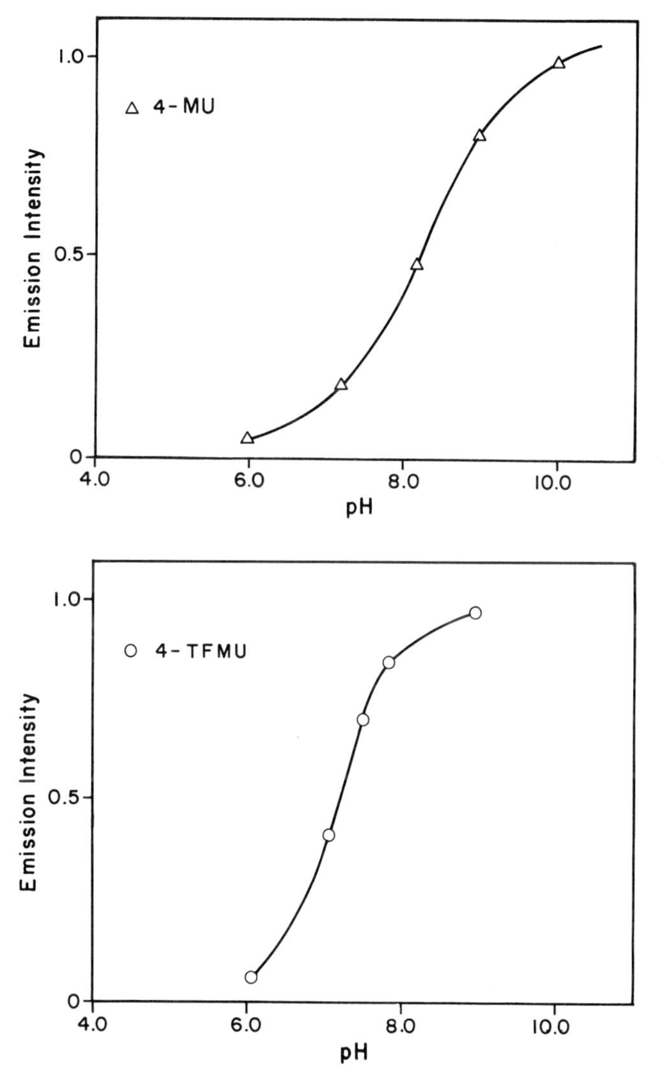

Fig. 4 The pH profiles for 4-methylumbelliferone (4-MU) and 4-trifluoromethylumbelliferone (4-TFMU) indicate the pK_a values for these fluorophores are 8.2 and 7.3, respectively.

Chapter 3). For protoplasts or other extracts that contain low levels of enzyme activity, time points are taken at longer periods (0, 16, and 40 h, for example) (Harkins *et al.*, 1990). Typically, the observed GUS activity remains linear over a long time course, and an initial time point is not necessary. In addition, the fluorescence intensity should be proportional to added enzyme (extract) as well.

For those laboratories lacking access to fluorescence instrumentation, a sensitive qualitative assay can be performed by placing samples on (or in) a long-wavelength (365 nm) ultraviolet (UV) light box and examining the production of the blue fluorescent product by sight. Positive and negative samples can quickly be distinguished, and even relative kinetic experiments can be measured by this means.

Resorufin β-D-Glucuronic Acid

Intrinsic fluorescence in cellular extracts, usually from large amounts of chlorophyll, or absorption from endogenous chromophores, which results in quenching, are serious problems in fluorescent analysis of β-glucuronidase activity in plant tissues. An alternative substrate to 4-MUG is resorufin β-D-glucuronic acid (Molecular Probes, Eugene, Ore., catalog number R-1161). This compound is cleaved by β-glucuronidase to produce resorufin (ex 571; em 584), which has a high extinction coefficient (56K), a high quantum efficiency value, and excitation, and emission wavelengths that avoid some of the background absorbance and fluorescence problems associated with endogenous compounds in a tissue extract.

The product, resorufin, additionally fluoresces maximally at neutral pH values, and is stable except in the presence of reducing agents (i.e., DTT), in which it readily undergoes reduction to a nonfluorescent product.

4-Trifluoromethylumbelliferyl β-D-Glucuronic Acid (4-TFMUG)

An alternate fluorescent substrate for β-glucuronidase detection is the 4-trifluoromethyl analog of 4-MUG, 4-trifluoromethylumbelliferyl β-D-glucuronide (Molecular Probes, Eugene, Ore., catalog number T-658), which releases trifluoromethylumbelliferone (pK_a 7.3; see Figures 4, 5, and 6) on cleavage by β-glucuronidase (Yegorov *et al.,* 1988).

This substrate has several improved fluorescence properties relative to 4-MUG. Due to the increased fluorescent yield at physiological pH levels, adjustment of the pH after enzymatic turnover is typically not necessary. Additionally, the longer-wavelength emission (ex 393; em 502) permits improved detection of fluorescence in the presence of

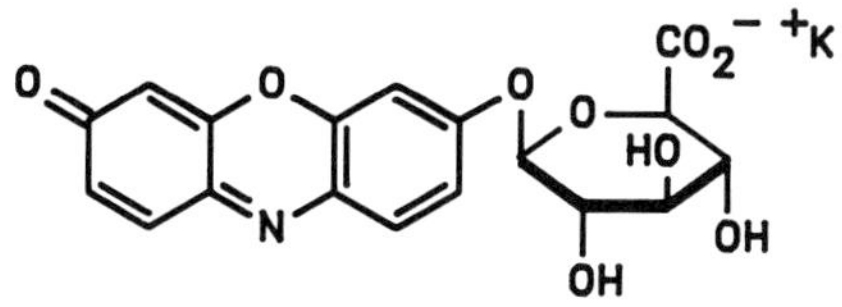

Fig. 5 Chemical structure of the fluorescent substrate resorufin β-D-glucuronic acid.

endogenous fluorescence in plant tissues (i.e., chlorophyll etc.). Finally, the relative enzymatic turnover rates for TFMU and MU glycosides are virtually identical (1.03 : 1.00 for the galactoside analogs), indicating no loss and, in fact, slightly improved substrate activity with this modification to the fluorophore, as would be expected with the addition of an electron-withdrawing group to the aglycone (Figure 7).

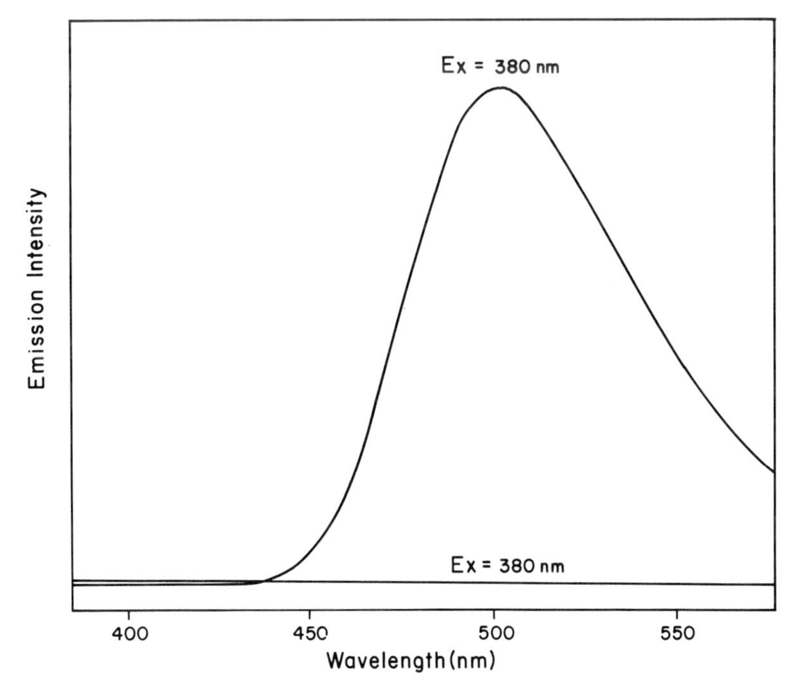

Fig. 6 Emission spectra for the substrate 4-trifluoromethylumbelliferyl β-D-glucuronide (4-TFMUG) and its enzymatic hydrolysis product 4-trifluoromethylumbelliferone (4-TFMU). The rate of enzymatic hydrolysis can be measured by the increase in fluorescence emission measured at 502 nm.

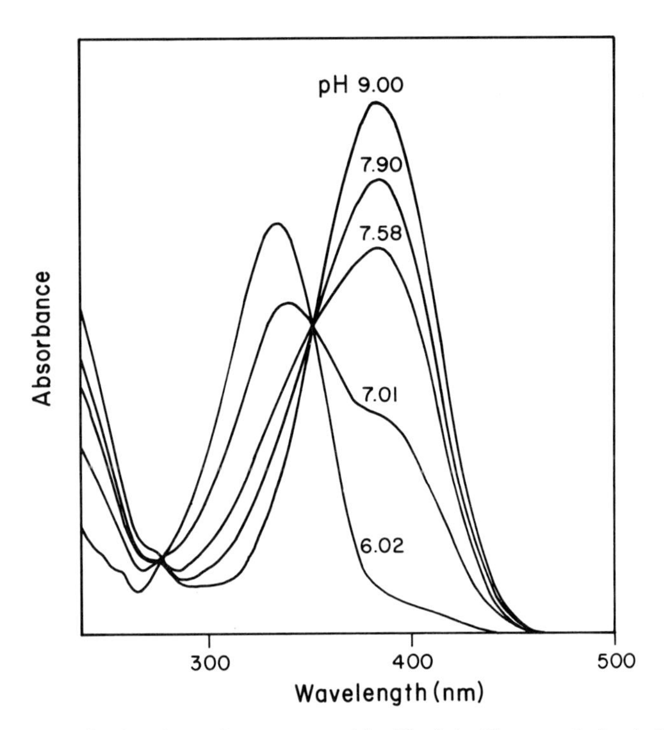

Fig. 7 Changes in the absorption spectra with pH of 4-trifluoromethylumbelliferone (4-TFMU) indicate an equilibrium between two species, namely, the phenol and phenoxide forms of the fluorophor. The fluorescent form of the dye (phenoxide form) has an absorption maximum at 393 nm.

Fig. 8 Chemical structure of the fluorescent histochemical substrate naphthol AS-BI glucuronide.

Fluorescein Mono-β-D-Glucuronide

Attempts to detect *in vivo* β-glucuronidase activity using fluorescein derivatives have met with limited success. A potential substrate for such analysis is the monoglucuronide of fluorescein (Molecular Probes, Eugene, Ore., catalog number F-1197). The intrinsic fluorescence of this substrate necessitates detection of enzymatic turnover using a front-face measurement method (Eisinger and Flores, 1979). Using front-face geometry versus right-angle geometry, one can overcome scattering and internal absorbance, including that from the fluorescent dye itself and from cellular chromophores. Accurate quantitation can therefore be obtained using high substrate concentrations, well above the K_m value for the substrate (Huang and Haugland, 1990).

Naphthol AS-BI β-D-Glucuronide

Naphthol AS-BI β-D-glucuronide (CalBioChem, San Diego, Calif., catalog number 477756; Sigma, St. Louis, Mo., catalog number N1875) is a substrate useful for histochemical determination of β-glucuronidase activity by a postcoupling technique (Figure 8) (Hayashi *et al.,* 1964; Fishman *et al.,* 1964). This substrate releases the fluorophore naphthol AS-BI, 6-bromo-2-hydroxy-3-naphthoyl-*O*-anisidine (ex 318/405, em 516), upon enzymatic hydrolysis. This intermediate is then reacted *in situ* with added hexazonium pararosanilin (or another aryl-diazo compound) to give a red precipitating azo dye through formation of an ortho-diazo derivative. The optimal staining reaction is obtained with 0.25 m*M* substrate and 1.8 m*M* diazo reagent at pH 5.2 in 20–30 min at 37°C. The brilliant red dye is visualized with a fluorescent microscope at the site of enzyme activity, often as discrete granules.

Properties of Substrates

An important property of fluorescent glucuronide substrates is their propensity to hydrolyze nonenzymatically. These substrates are quite stable, however, if stored in a solid form, desiccated, and kept cold (below 0°C). Stock solutions of the substrates should be prepared by first dissolving the substrate in a minimum amount of an appropriate organic solvent [dimethyl sulfoxide (DMSO) or dimethylformamide (DMF) is often a good choice]. These solutions are then diluted to the desired concentration for analysis by adding an appropriate buffer [sodium phosphate (PBS), or Tris, pH 7]. Since addition of organic solvents can be detrimental in some cellular assays, the substrate can be alternately dissolved directly in the working buffer by gentle agitation, vortex mixing, or ultrasonication. Heating of substrate solutions is not recommended, since this will increase the rate of hydrolysis. Long periods of ultrasonication should also be avoided, since these often result in significant heating of the samples. Stock solutions should be kept in an ice bath during use and frozen after use. Thawing of these solutions should be performed with the same delicacy as noted above. These working solutions are often stable for weeks if the above precautions are observed. A simple analysis of these solutions, either by UV absorbance analysis or by thin-layer chromatography [TLC silica-gel plates (Kieselgel 60), 0.2 mm thickness], can be used to reveal trace decomposition. Refer to the data in Table 1 for such analyses.

Background fluorescence can sometimes be the result of high levels of endogenous β-glucuronidase enzyme, or enzyme that is inadvertently added during the production of protoplasts using commercially available enzyme mixtures (Rhozyme). In addition, a combination of reagents used in culture media (Fe_2EDTA in combination with DTT) has greatly stimulated nonspecific hydrolysis of 4-MUG and similar glycosides. Finally, long incubation times will usually increase the

Table 1

Analytical Data for Selected Fluorophores and Fluorescent Substrates[a]

	TLC data		UV data	
Compound	R_f	Solvent[b]	ε (10^{-3})	λ_{max} (nm)
4-MU	0.78	A	19	360 (pH 9)
4-MUG	0.63	B	15	317
4-TFMU	0.57	A	16	393 (pH 9)
4-TFMUG	0.34	C	11	325
FMG	0.47	D	26	450
Fluorescein	0.24	A	86	490 (pH 9.5)
X-GlcU	0.55	E	4.7	292 (MeOH)
Resorufin-GlcU	0.33	B	14	468
Resorufin	0.55	F	56	571 (pH 9)

[a] The molar concentration of a pure analyte can be determined spectrophotometrically by measuring the absorbance at λ_{max} and applying the Beer–Lambert law; λ_{max} was determined in aqueous solution except where noted. For optimum fluorescence, the pH was raised to 9 or above with KOH. TLC was run on SiO_2 plates with the solvents listed.

[b] Solvent systems: A = 19 : 1 chloroform : methanol; B = 7 : 1 : 1 : 1 ethyl acetate : methanol : water : acetic acid; C = 9 : 1 isopropanol : water; D = 3 : 1 : 1 butanol : methanol : water; E = 8 : 2 : 1 butanol : acetic acid : water; F = 8 : 2 chloroform : methanol.

amount of nonspecific hydrolysis of fluorogenic substrates. Since the detection levels of the fluorescent products formed are quite sensitive, long assays should be avoided.

Working concentrations for fluorescent substrate stock solutions are typically in the range of 100 μM to 2.0 mM (~0.5 mg/ml).

A convenient glucuronidase inhibitor saccharo-lactone (saccharic acid, 1,4-lactone, Sigma, St. Louis, Mo., catalog number S-0375) can be used to delineate weakly fluorescent cells/assays. Since the inhibitor is base labile, test reactions should be performed at pH 6.

Synthesis of Substrates

The original isolation of β-glucuronic acid conjugates from rabbit and human urine following oral administration often involves complex extraction and column chromatography techniques, and the amount that can be extracted is generally small (Chen *et al.*, 1980). An exception is the isolation of 4-methylumbelliferyl glucuronide (Mead *et al.*, 1955), in which a 20–25% isolated yield can be obtained from a 2.5-g administered dose.

In general, however, synthetic pathways to these fluorogenic substrates provide the simplest means of obtaining the desired compounds. Typically, Koenigs–Knorr methodology is employed (Matsunaga *et al.*, 1984; Courtin-Duchateau and Veyrieres, 1978), in which a phenol containing fluorophore and acetobromoglucuronide methyl ester are combined under anhydrous conditions in the presence of a silver or mercury catalyst to give the fully protected β-glucuronide product. Removal of the acetate protecting groups and saponification of the methyl ester provides the free glucuronide product, often in fair to good yield. Purity of the final substrate can be determined by measurement of the UV spectrum (extinction coefficient), which should be insensitive to pH.

Prospective for Future Substrates

Currently, several laboratories are involved in a search for improved fluorescent substrates for single cell detection of β-glucuronidase activity *in vivo,* which could be used in conjunction with fluorescence-activated cell sorting (FACS) techniques. It has been possible to adapt the techniques of flow cytometry and fluorescence-activated cell sorting for the analysis of plant protoplasts while maintaining their viability (Galbraith, 1989). Attempts to use the substrate fluorescein di-β-D-glucuronide for such an application were unsuccessful (D. Galbraith, personal communication), apparently because the substrate was impermeant to the protoplast cell membrane. A fluorescent substrate with both improved cell uptake and retention properties after enzymatic hydrolysis, through precipitation, intracellular reaction, or specific binding to intracellular components, will represent the next generation of fluorometric GUS substrates for this application.

In addition, recent developments in the laboratories of A. P. Schaap at Wayne State University and Irena Bronstein at Tropix, Inc. (Bronstein and McGrath, 1989), indicate the possibility of utilizing chemiluminescent substrates for detection of GUS activity *in vivo*. In such an assay, a dioxetane chemiluminescent substrate would emit energy in the wavelength of visible light upon enzymatic cleavage. Despite some difficulties, chemiluminescence could offer several advantages over other analytical labeling techniques, including high sensitivity with low background levels.

Conclusion

Analysis of GUS activity in plants by use of fluorometric substrates provides the most sensitive as well as a convenient method of quantitating various gene fusion parameters involved in successful production of transgenic plants. The most commonly used fluorescent substrate is 4-methylumbelliferyl β-D-glucuronide. For viable cell assays, the histochemical substrate X-GlcU is the substrate of choice. In addition, improved analytical techniques and new substrates are now available and are continuously being updated. Future research will most likely focus on routine analysis of GUS gene fusions using fluorometric substrates in intact live cells (for FACS analysis).

Acknowledgments

The author would like to thank David Galbraith for his helpful comments and suggestions, John Mezner and Peter Hewitt for help in producing absorbance and fluorescence spectra, Nan Minchow for the preparation of technical illustrations, and Alisa Naleway for her patient proofreading and editing of this manuscript. Funding from NIH grant R44 GM38987-02 is gratefully acknowledged.

References

Bronstein, I., and McGrath, P. (1989). Chemiluminescence lights up. *Nature* 338, 599–600.

Chen, R. F. (1968). Fluorescent pH indicator. Spectral changes of 4-methylumbelliferone. *Anal. Lett.* 1(7), 423–428.

Chen, S. C., Nakamura, H., and Tamura, Z. (1980). Studies on the metabolites of fluorescein in rabbit and human urine. *Chem. Pharm. Bull.* 28(5), 1403–1407.

Courtin-Duchateau, M. C., and Veyrieres, A. (1978). Synthesis of 4-methylumbelliferyl 1,2-*cis*-glycosides. *Carbohydrate Res.* 65, 23–33.

Eisinger, J., and Flores, J. (1978). Front-face fluorometry of liquid samples. *Anal. Biochem.* 94, 15–21.

Fishman, W. H., Nakajima, Y., Anstiss, C., and Green, S. (1964). Naphthol AS-BI β-D-glucosiduronic acid; Its synthesis and suitability as a substrate for β-glucuronidase. *J. Histochem. Cytochem.* 12, 298–305.

Galbraith, D. W. (1989). XVIII Flow cytometric analysis and sorting of somatic hybrid and transformed protoplasts. *In* "Biotechnology in Agriculture and Forestry," Vol. 9, "Plant Protoplasts and Genetic Engineering II" (ed. Y. P. S. Bajaj), Springer-Verlag, Berlin, pp. 304–327.

Gersson, D. F. (1982). Determination of intracellular pH changes in lymphocytes with 4-methylumbelliferone by flow microfluorometry. *In* "Intracellular pH: Its Measurement, Regulation, and Utilization in Cellular Functions." Alan R. Liss, New York, pp. 125–133.

Harkins, K. R., Jefferson, R. A., Kavanagh, T. A., Bevan, M. W., and Galbraith, D. W. (1990). Expression of photosynthesis-related gene fusions is restricted to defined cell types in transgenic plants and in transfected protoplasts. *Proc. Natl. Acad. Sci. USA,* **87,** 816–820.

Hayashi, M., Nakajima, Y., Fishman, W. H. (1964). The cytologic demonstration of β-glucuronidase employing naphthol AS-BI glucuronide and hexazonium pararosanilin; A preliminary report. *J. Histochem. Cytochem.* 12, 293–297.

Hu, C.-Y., Chee, P. P., Chesney, R. H., Zhou, J. H., Miller, P. D., and O'Brien, W. T. (1990). GUS-like activities in seed plants. *Plant Cell Reports* 9, 1–5.

Huang, Z. (1991). Kinetic assay of fluorescein mono-β-D-galactoside hydrolysis by β-galactosidase: A front face measurement method for strongly absorbing fluorogenic substrates. *Biochemistry,* **30,** 8530–8534.

Jefferson, R. A. (1987). Assaying chimeric genes in plants: The GUS gene fusion system. *Plant Mol. Biol. Rep.* 5, 387.

Jefferson, R. A., Kavanagh, T. A., and Bevan, M. W. (1987). GUS fusions: β-Glucuronidase as a sensitive and versitile gene marker in higher plants. *EMBO J.* 6, 3901–3907.

Lojda, C. (1970). Indigogenic methods for glycosidases II. An improved method for β-D-galactosidase and its application to localization studies of the enzymes in the intestine and in other tissues. *Histochemie* 23, 266–288.

Matsunaga, I., Nagataki, S., and Tamura, Z. (1984). Synthesis of fluorescein monoglucuronide. *Chem. Pharm. Bull.* 32(7), 2832–2835.

McCabe, D. E., Swain, W. F., Martinell, B. J., and Christou, P. (1988). Stable transformation of soybean (glycine max) by particle acceleration. *Biotechnology* 6, 923–926.

Mead, J., Smith, J., and Williams, R.T. (1955). Studies in detoxication 67. The biosynthesis of the glucuronides of umbelliferone and 4-methylumbelliferone and their use in fluorometric determination of β-glucuronidase. *Biochem. J.* 61, 569.

Pearson, B., Andrews, M., and Grose, F. (1961). *Proc. Soc. Exp. Biol.* 108, 619–623.

Schulz, M., and Weissenbock, G. (1987). Partial purification and characterization of a luteolintriglucuronide-specific β-glucuronidase from rye primary leaves (*Secale cereale*). *Phytochemistry* 26, 933.

Sood, P. P. (1980). Histoenzymological compartmentation of β-glucuronidase in the germinating pollen grains of *Portulaco grandiflora*. *Biologia Plantarum (Praha)* 22, 124.

Tomasic, J., and Keglevic, D. (1973). The kinetics of hydrolysis of synthetic glucuronic esters and glucuronic ethers by bovine liver and *Esherichia coli* β-Glucuronidase. *Biochem. J.*, 133, 789.

Yegorov, A. M., Markaryan, A. N., Vozniy, Y. V., Cherednikova, T. V., Demcheva, M. V., and Berezin, I.V. (1988). 4-Trifluoromethylumbelliferyl β-D-galactopyranoside: Its synthesis and application as the fluorogenic substrate of β-galactosidase *E. coli* for screening monoclonal antibodies by immunosorbent ELISA. *Anal. Lett.* 21(2), 193–209.

5 Automated Preparation of Plant Samples for Enzymatic Analysis

Thomas B. Brumback, Jr.
Pioneer Hi-Bred International, Inc.
Johnston, Iowa

The increasing use of recombinant DNA techniques in plant genetic research has expanded the need for rapid procedures to extract proteins from plant tissues. The detection and analysis of proteins expressed by cloned genes are usually preceded by some extraction method to liberate the proteins from cellular material. Often, the most laborious step in most extraction methods involves the mechanical disruption of the cellular material. Whereas bacteria, fungi, and mammalian cells can be lysed by a variety of techniques including sonication, hypertonic solutions, mild mechanical vortexing, and enzymatic lysis (Sambrook *et al.*, 1989), plant tissues usually require more vigorous mechanical means to disrupt cell walls. Typically, such disruption is achieved by common pestle and tube grinding methods, grinding with sand or glass beads, homogenizers, or by more esoteric methods such as leaf presses or decompression homogenization. Many of these methods are difficult to automate in a laboratory environment.

Our manual procedure for grinding small amounts of leaf and callus tissue used disposable pestles (Kontes Glass, Vineland, N.J.) that fit into Eppendorf microcentrifuge tubes. We have modified this pestle-and-tube grinding method and have developed a robotic system for automating the extraction process.

GUS Protocols: Using the GUS Gene as a Reporter of Gene Expression

System Design Criteria

Few laboratory procedures are designed with automation as a major goal. Manual procedures are usually optimized for humans and rarely are suited to direct automation. Procedures that would be nearly impossible for a human to perform can be efficiently performed by robots. To automate any protocol, the number of steps must be minimized and optimized for automation. The initial design criteria for our protocol are outlined in Table 1. From these design criteria and numerous manual methods routinely used in the laboratories, a robotic protocol was developed (Table 2). After the extraction is complete, the protein concentration of the supernatant is determined by the Bradford method (Bradford, 1976) prior to running one or more assays for specific proteins such as β-glucuronidase (GUS) (Rao and Flynn, 1990).

Robotic System

A custom robotic system was developed by Bohdan Automation (Chicago, Ill.) according to our design objectives. The basic unit is microcomputer controlled and consists of two overhead arms mounted on a 4 ft by 5 ft table (Figure 1). The arms move sample tubes among four

Table 1

Design Requirements of Automated Sample Preparation System

Design requirement	Reason/advantage
1. Small sample size.	Minimize reagent use and reduce time required to obtain sufficient sample for analysis.
2. High throughput of at least 100 samples per hour.	This is the minimal throughput required for large screening projects.
3. Use standard labware (microcentrifuge tubes and microassay plates).	Allows integration with other instruments and is readily available.
4. Minimize and simplify protocol steps.	Reduces design complexity and system costs.
5. Cool samples during each step.	Minimizes protein degradation.

Table 2

Robotic Protocol for Grinding Callus and Leaf Tissue

1. Place sample (5 mg leaf punch or 30–40 mg callus) into 1.5 ml Sarstedt screw-cap microcentrifuge tube.
2. Remove tubes from sample rack and dispense 170 μl cold lysis buffer into tubes.
   GUS Lysis Buffer:
   50 m*M* sodium phosphate buffer, pH 7.0
   1 m*M* EDTA
   10 m*M* β-mercaptoethanol
3. Grind samples for 20 s.
4. Centrifuge at 8000 × g for 4 min.
5. Return microcentrifuge tubes to sample rack and pipet 85 μl supernatant into microassay plate.

different workstations to (1) deliver buffer, (2) grind samples, (3) centrifuge samples, and (4) pipet supernatant into a microassay plate. Located beneath the table are support units for cooling, uninterruptible power, waste collection, and compressed air supply. A schematic drawing of the system is shown in Figure 2. Details and design features of the system are described elsewhere (Brumback, 1991).

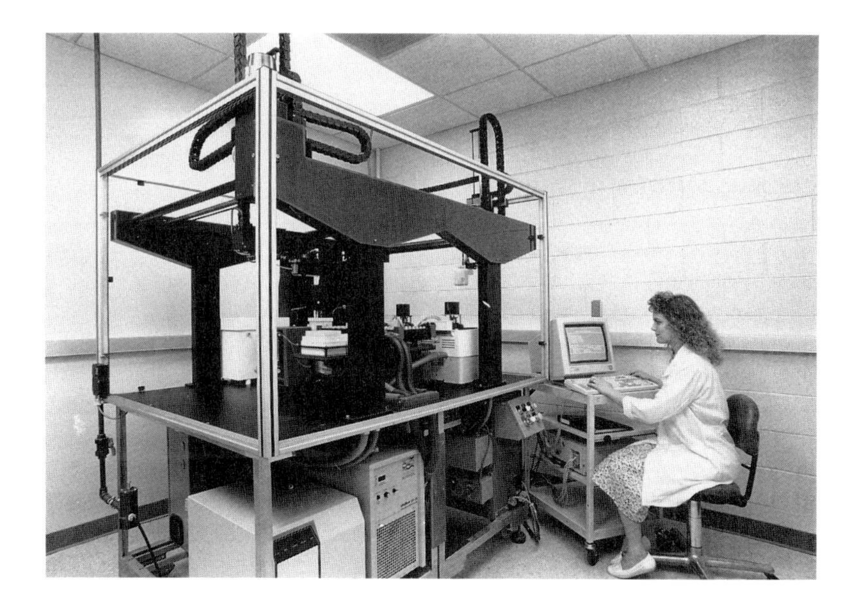

Fig. 1 Robotic system for plant protein extraction.

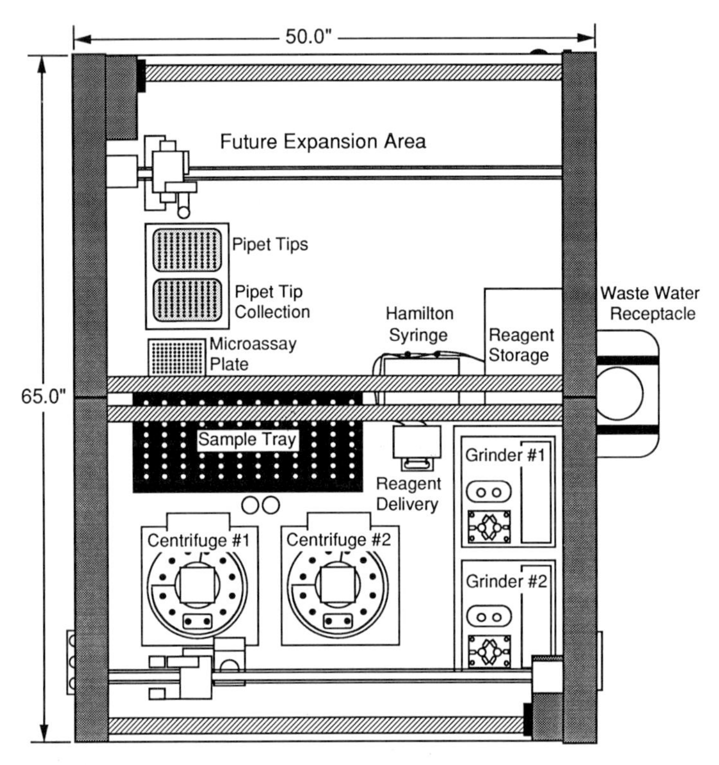

Fig. 2 Schematic of robotic system for plant protein extraction.

System Operation

Prior to operation, callus or leaf punch samples are manually placed in 1.5-ml screw-cap microcentrifuge tubes (Sarstedt, Princeton, N.J.). The tubes are held in an ice-cooled plastic sample tray capable of holding up to 96 tubes. Typically, a sample tray can be filled in 15–20 min. A loaded sample tray is then placed on the refrigerated sample rack in the robot. The operator then fills a buffer reservoir with the appropriate lysis buffer and primes the dispenser. From a menu option on the microcomputer control program, the operator selects an appropriate protocol from a list of possible procedures and runs a program to start processing samples.

Using a specialized gripper, the robotic arm removes sample tubes from the sample rack two at a time to improve throughput (Figure 3). The samples are moved to a dispensing/cooling tower connected to a Hamilton MicroLab programmable dispenser. The dispenser delivers a specified amount of chilled buffer to each of the sample tubes. The cooling tower cools the buffer to 2°C just prior to delivery to help maintain the low temperatures needed to minimize protein degradation.

After buffer addition, the samples are moved to a custom grinding unit (Figure 4), which uses two stainless steel pestles to grind the samples directly in the microcentrifuge tubes. The pestles are machined to closely fit the inside dimensions of the microcentrifuge tubes. While the pestles rotate, they occasionally move up and down to ensure mixing and improve grinding. When grinding is complete, the pestles are washed with a jet of distilled water, and any adhering droplets are removed by a small jet of compressed air.

Tubes containing the ground samples are then moved to a Savant high-speed centrifuge and spun at 8000 × g for 4 min. Using a swinging-bucket rotor, the centrifuges are capable of spinning 12 samples at a time. After a batch of 12 samples has been centrifuged, the tubes are

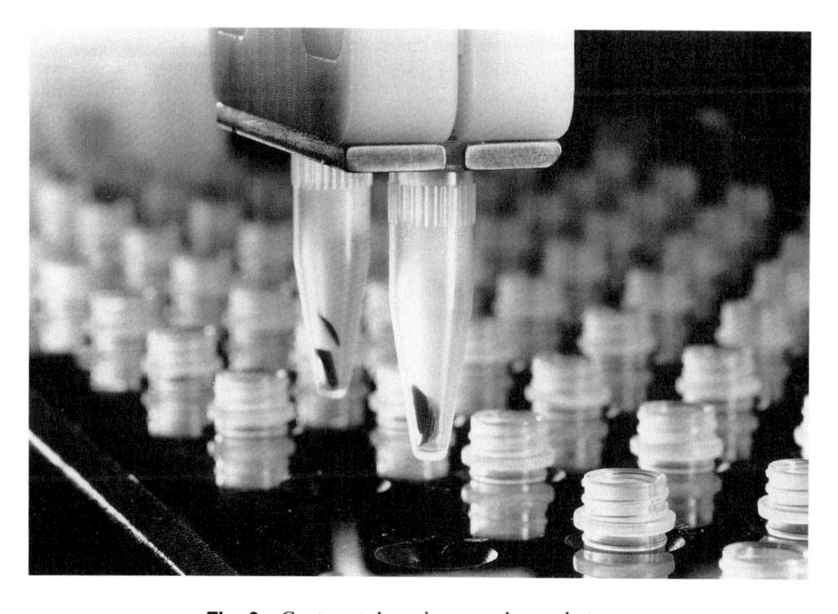

Fig. 3 Custom tube gripper and sample tray.

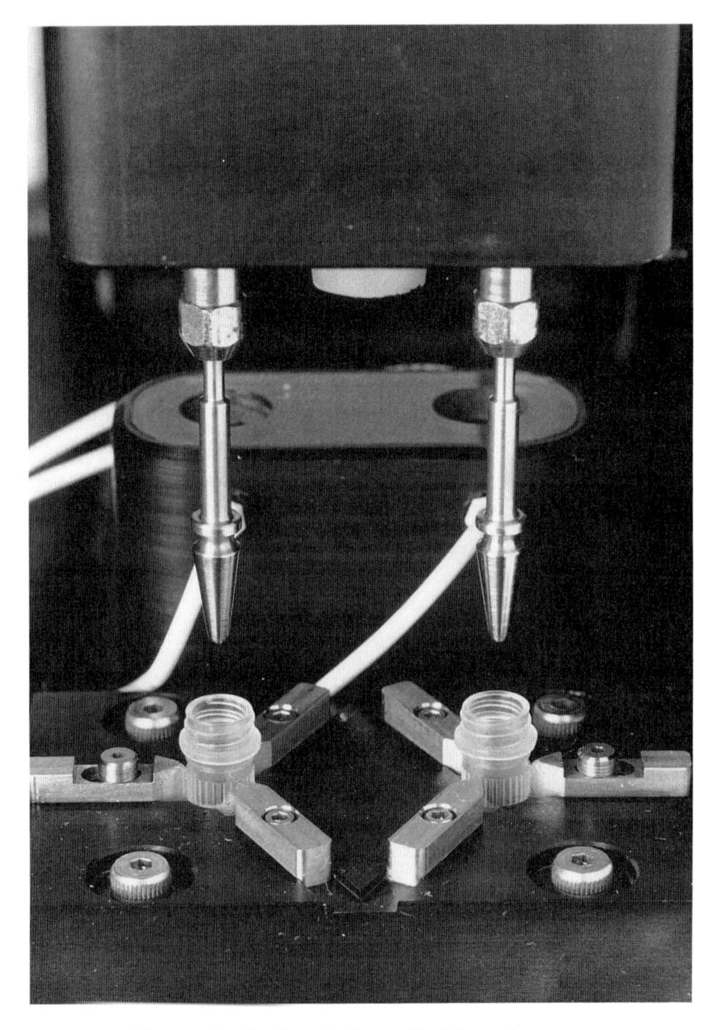

Fig. 4 Dual-tube grinding unit with wash cups.

moved to the sample rack where a second robotic arm pipets the supernatant into a microassay plate to complete the sample preparation process. Disposable pipet tips are used to prevent sample cross-contamination. The process repeats until all samples have been processed.

Methods Development Software

The heart of the software that controls robotic operations is a methods editor. The extraction procedure can be applied to an infinite variety of samples, many requiring a unique approach to extraction. The methods editor software allows any number of variations in methodology to be used. It enables the user to enter and store parameters suited to the particular group of samples to be analyzed. Examples of such parameters are buffer name, dispense volume, centrifuge and grinding times, and delivery volume. The editor presents a series of on-screen pages for the user to fill out and allows the user to create, edit, delete, and select methods to execute (Figure 5). The use of methods and the methods editor simplifies procedures for the user, since routine extractions can be run by simply selecting and executing the appropriate stored method.

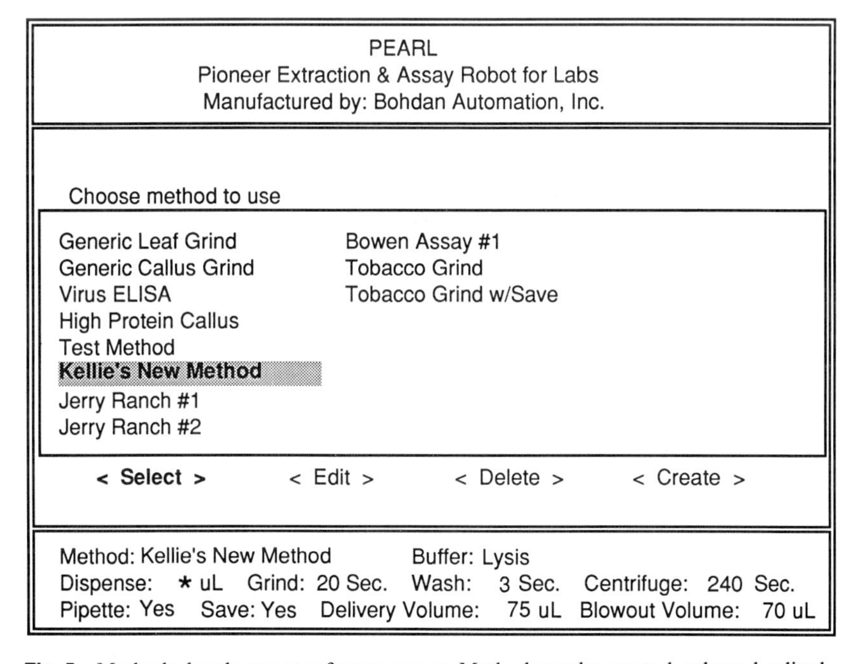

Fig. 5 Methods development software screen. Methods can be created, selected, edited, or deleted. Method parameters are displayed at the bottom.

Validation

Since the manual procedures were modified significantly during the automation process, efforts were required to validate the new procedures and to evaluate the efficiency of the robotic system. Several experiments were conducted to evaluate grinding efficiency, cross-contamination, and overall performance.

A comparison of manual and robotic procedures was made by analyzing the protein extraction from tobacco leaf punches (Table 3). The Bradford protein assay was used to determine protein content of each of 24 samples extracted by the manual and robotic methods. The robotic system extracted significantly more protein and produced nearly 50% less variability among samples than a similar manual grinding method. Apparently, the stainless steel grinding pestles are able to exert more force and have larger grinding surfaces than the Kontes pellet-pestle tubes used in the manual method. The result is improved extraction efficiency.

The use of dual grinders, each containing dual pestles, creates the potential for differential protein extraction among these units. Since subsequent assays are usually based on the amount of protein extracted, only extreme variation in extraction efficiency among grinders and pestles would be deleterious. Protein extraction tests showed no significant differences among grinders and pestles (Table 4). Coefficients of variation ranged from 14 to 20%. Since the system was not designed for exhaustive protein extraction, no tests of percent protein recovery were made.

The fixed-pestle design of the grinders requires that the pestles be thoroughly cleaned after each sample is ground. Though jets of distilled

Table 3

Comparison of Manual and Robotic Grinding Methods on Protein Extraction from Tobacco Leaf Punches

Grinding method	n	Protein (μg/μl)	CV
Manual	24	1.18 ± .18	15%
Robotic	24	1.33 ± .10[a]	8%

[a] Denotes significant difference between procedures at the 0.01 probability level. One leaf punch (~5 mg) of tissue was used. CV, coefficient of variation.

Table 4

Comparison of Different Grinders and Pestles on Protein Extraction from Tobacco Leaf Punches[a]

	Protein extracted ($\mu g/\mu l$)			
	Pestle 1	Pestle 2	Mean	CV
Grinder 1	1.19 ± .13	1.11 ± .18	1.15 ± .16	14%
Grinder 2	1.24 ± .27	1.14 ± .21	1.19 ± .24	20%
Mean	1.21 ± 21	1.13 ± 19		
CV	17%	17%		

[a] At P = .05, no significant differences were detected among grinders or pestles (n = 12). CV, Coefficient of variation. One leaf punch (~5 mg) of tissue was used.

water and air are used to rinse and dry the pestles, the potential exists for contamination or carryover between samples. Preliminary experiments showed (Table 5) that protein carryover was minimal, occurring in only one sample out of 28. With large sample amounts and buffer volumes in excess of 250 ml, the likelihood of carryover increases. In other experiments in which over 300 samples were tested, we detected only two isolated events of protein carryover and these events occurred with high sample volumes. Lower sample volumes and longer wash times will reduce the possibility of contamination from protein carryover. Other contamination tests using a more sensitive ELISA procedure to detect alfalfa mosaic virus (AMV) coat protein contamination in alfalfa have shown no contamination or carryover among samples.

System Performance

The system has been used to extract proteins from a wide variety of plant tissues including tobacco, corn, sunflower, soybean, and alfalfa. Approximately 70% of all samples processed to date have been callus tissues. In its current configuration, the unit is capable of processing 96 samples in 45 min. Usually, about 1 h is required for a 96-sample batch since the operator must reload the system with new pipet tips, a microassay plate, and new samples. Thus, about 768 samples can be processed in an 8-h day. During the 13 months since the unit was

Table 5

Protein Carryover during Grinding of Different Corn (*Zea mays*) Tissues and Protein Concentrations[a]

Sample	Protein concentration ($\mu g/\mu l$)	
	Sample	Adjacent tubes
1 Leaf punch	0.613 ± 0.131	0.001 ± 0.002
1 Leaf punch + 5 μg BSA	0.730 ± 0.063	0.002 ± 0.002
2 Leaf punches	1.185 ± 0.038	0.001 ± 0.001
2 Leaf punches + 5 μg BSA	1.008 ± 0.070	0.001 ± 0.002
2 Leaf punches + 50 μg BSA	2.427 ± 0.072	0.003 ± 0.003
25 mg callus	0.997 ± 0.112	0.007 ± 0.001
25 mg callus + 5 μg BSA	0.837 ± 0.100	0.003 ± 0.004
50 mg callus	1.487 ± 0.175	0.005 ± 0.004
50 mg callus + 5 μg BSA	1.218 ± 0.113	0.006 ± 0.005
150 mg callus	2.056 ± 0.171	0.005 ± 0.004
150 mg callus + 5 μg BSA	2.021 ± 0.127	0.052 ± 0.071[b]
150 mg callus + 50 μg BSA	2.489 ± 0.113	0.002 ± 0.003
5 μg BSA	0.463 ± 0.152	0.001 ± 0.001
50 μg BSA	1.926 ± 0.171	0.003 ± 0.002
Blank	0.004 ± 0.008	0.002 ± 0.002

[a] BSA (bovine serum albumin) was added as 50 μl of appropriate concentration to yield specified amount. Total sample volume was 150 μl buffer + sample + added BSA. Leaf punches were approximately 5 mg. Samples containing protein and/or plant tissue were ground two at a time. Following each grind, tubes containing only buffer (adjacent tubes) were ground and were used to detect carryover.

[b] Denotes significant protein carryover.

brought into production, about 14,000 samples have been processed with a maximum daily output of 640 samples. These samples have been used successfully in a variety of assays for total protein (Bradford, 1976), GUS (Rao and Flynn, 1990), BAR (the BAR gene confers resistance to the commercial herbicide BASTA, which interferes with glutamine synthetase; DeBock *et al.*, 1987), AMV coat protein (unpublished data), and NPTII (unpublished data).

The automation of the extraction process represents the first step in the overall automation of enzymatic assays. In the future, we plan to integrate a Beckman Biomek 1000 automated laboratory workstation to perform assays directly on the robotic table. The Biomek will be integrated with our system software for control and will provide the capabilities to do a variety of assays and protocols. In addition, we plan to convert the system from a batch-mode operation (i.e., processing one set of samples before reloading) to a multiple-batch mode (i.e., processing multiple sample sets before reloading) to allow continuous operation without operator intervention. With the integrated system, users will be able to load samples for processing and return when completed analyses are ready.

Acknowledgments

I would like to thank Lyndon Schroeder and Teresa Beghtol for technical advice and operation of the robot, Kellie Winter and Pam Flynn for the experimental data, Guru Rao and Ben Bowen for helpful comments on the manuscript, and Kristy Mckenna for preparation of the manuscript.

References

Bradford, M. M. (1976). A rapid and sensitive method for the quantitation of microgram quantities of protein utilizing the principle of protein-dye binding. *Anal. Biochem.* 72:248–254.

Brumback, T. B., Jr. (1991). Automated robotic extraction of proteins from plant tissue samples. *In* "Advances in Laboratory Automation Robotics" (J. R. Strimaitis and James N. Little, eds.), Vol. 7, Zymark Corp, Hopkinton, Mass.

DeBlock, M., Botterman, J., Vandewiele, M., Dockx, J., Thoen, C., Van Montagu, M., and Leemans, J. (1987). Engineering herbicide resistance in plants by expression of a detoxifying enzyme. *EMBO J.* 6:2513–2518.

Rao, A.G. and Flynn, P., (1990). A quantitative assay for β-D glucuronidase (GUS) using microtiter plates. *BioTechniques* 8(1):38–40.

Sambrook, J., Fritsch, E. F., and Maniatis, T. (1989). "Molecular Cloning: A Laboratory Manual," 2nd ed. Cold Spring Harbor Laboratory Press, Cold Spring Harbor, N.Y.

6 Microtiter Plate-Based Assay for β-D-Glucuronidase: A Quantitative Approach

A. Gururaj Rao and Pamela Flynn
Pioneer Hi-Bred International
Johnston, Iowa

The gene encoding β-D-glucuronidase (GUS) has been widely used as a reporter gene in plant transformation systems following the report by Jefferson *et al.* (1986) describing a gene-fusion system with the GUS gene from *Escherichia coli*. The bacterial enzyme has an M_r of ~70,000 daltons and catalyzes the cleavage of a large number of glucuronides. The enzyme lends itself to easy detection through a variety of substrates, many of which were developed in the 1950s and 1960s during the course of research on GUS from mammalian sources. These included substrates for spectrophotometric and fluorometric measurements and also histochemical staining. However, fluorescence measurement is many orders of magnitude more sensitive than absorption measurement. The most widely used fluorogenic substrate, 4-methylumbelliferyl-β-D-glucuronide (MUG), was, in fact, synthesized by Mead *et al.* in 1955. They observed that although umbelliferone fluoresced strongly in UV light at pH 9–10, its conjugates showed little or no fluorescence. They therefore synthesized the glucuronide of the cheaper 4-methylumbelliferone as a very sensitive substrate for the fluorimetric determination of GUS activity in animal tissues. The other fluorescent GUS substrates that have been synthesized (Figure 1) and are commercially available are 4-trifluoromethylumbelliferyl-β-D-glucuronide (TUG) and Resorufin-β-D-glucuronide (REG); see Naleway (Chapter 4) for a complete description of the substrates. In this chapter we describe conditions for measuring GUS activity in plants in microtiter plates using the three substrates mentioned above.

GUS Protocols: Using the GUS Gene as a Reporter of Gene Expression

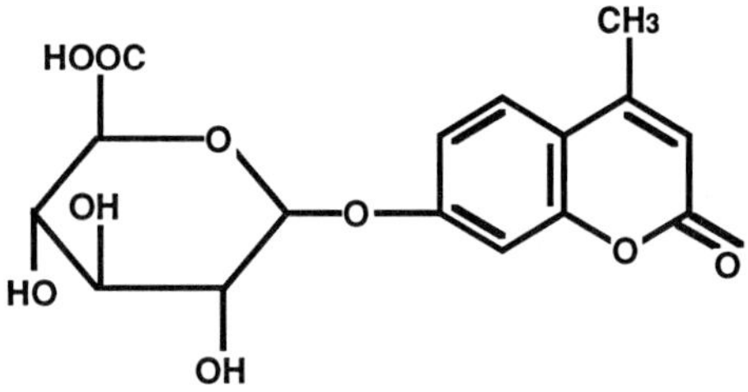

MUG

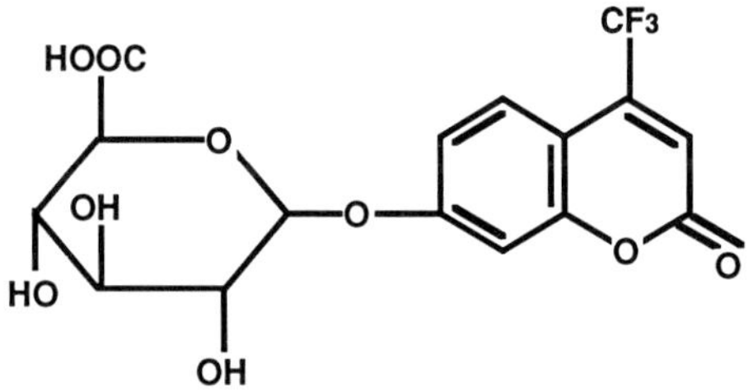

TUG

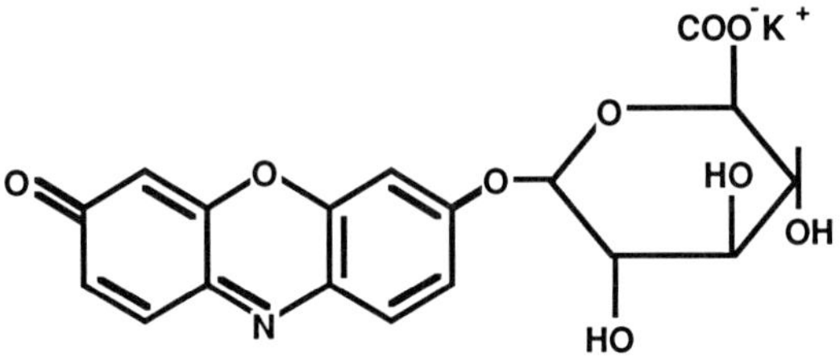

REG

Fig. 1 The structure of the substrates MUG, TUG, and REG (top to bottom).

Instrumentation

The basis of the GUS assay is the detection of fluorescence from the intensely fluorescent product of enzymatic hydrolysis of the substrate. For example 4-methylumbelliferone (4-MeU) is the fluorescent product of the substrate MUG. The accuracy of the assay depends on the sensitivity of the instrument used to detect the fluorescence.There are two instruments that are available in the market today with the ability to read fluorescence from microtiter plates.

One is the Titertek Fluoroskan II from Flow Laboratories (Mclean, Va.). In this instrument the choice of excitation and emission wavelengths is determined by four fixed excitation filters (355, 485, 544, and 584 nm) and four emission filters (460, 538, 590, and 612 nm). The fluorescence of 4-MeU is measured with an excitation at 355 nm and emission at 460 nm. The excitation light source is a xenon lamp and the light is transmitted through a quartz fiber optics cable that is positioned above the microtiter plate. The emitted light passes through another fiber optics cable located above the plate and is detected by a photomultiplier tube, which can detect light in the 300–700 nm range. The sensitivity of Fluoroskan II is illustrated by its ability to detect low levels of 4-MeU fluorescence in the 3 nM to 25 μM range (Figure 2).

The second instrument is the model LS 50 Luminescence Spectrometer from Perkin-Elmer. This is a computer-controlled, menu-driven instrument equipped with a special xenon flash tube and excitation and emission reflection grating monochromators with continuous

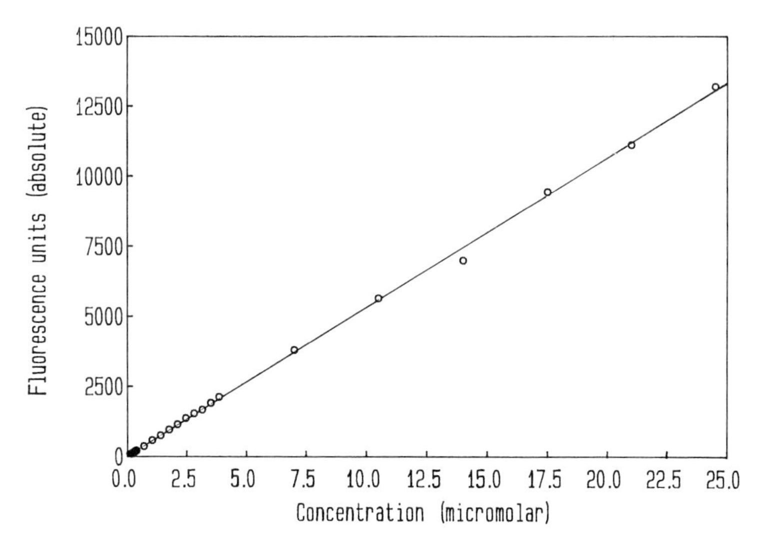

Fig. 2 Absolute fluorescence of 4-MeU in 0.2 M Na_2CO_3 as a function of concentration. A stock solution of 4-MeU (Sigma Chemicals, St. Louis, Mo.) was made and the concentration was determined using a molar extinction coefficient of 19,000 at 360 nm. Solutions of up to 25 μM were prepared by appropriate dilution of the stock in 0.2 M Na_2CO_3 and 200-μl aliquots were used in fluorescence measurements (from Rao and Flynn, 1990).

scanning capability in the 200–800 nm range. The microtiter plate reader comes as an accessory. Here also the excitation and emission light are transmitted through fiber optics cables located above the plate. However, unlike the Fluoroskan, where measurements are limited to a few wavelengths, excellent sensitivity is achieved for a wide range of fluorophors over the range from 300–720 nm.

The Determination of GUS Activity

The keys to obtaining quantitative information on GUS activity in transformed plant samples are (1) the establishment of a standard curve showing the relationship between concentration of pure enzyme and activity and (2) a knowledge of the influence of tissue extracts on the activity.

Preparation of Standard Curve

1. Prepare a stock solution of GUS [type VII-A from *E. coli* (Sigma Chemicals) in 0.1 *M* sodium phosphate buffer, pH 7.0, and determine the protein concentration by the method of Bradford (1976). Generally, adding 1 ml of buffer per vial results in a 0.03–0.1 mg/ml solution. Make a dilution of this in lysis buffer (50 m*M* sodium phosphate, pH 7.0, 10 m*M* EDTA, 0.1% Triton X-100, 0.1% sarkosyl, and 10 m*M* β-mercaptoethanol) to give a working solution of 0.1 ng/μl.
2. Dispense an appropriate volume of lysis buffer into the wells of an opaque microtiter plate (Perkin-Elmer) specifically meant for fluorescence measurements. This comes in a choice of either black or white but the latter is preferable. Note that the final volume of the reaction mix after addition of all the reagents is 50 μl.
3. From the working solution of GUS prepared above, add to each well 0.2–2 ng of GUS. Ensure that there are a few wells into which no enzyme is dispensed, to serve as a blank.
4. Initiate the reaction by the addition of 5 μl of the substrate, MUG (Sigma), from a 10 m*M* stock solution in 0.1 *M* sodium phosphate buffer, pH 6.5 (can be stored at −20°C for several months).
5. Cover the plate and incubate at 37°C for time periods ranging

from 10 to 60 min. This can be done either in the Fluoroskan itself, where there is an option to warm the plate to 37°C, or externally in an incubator.
6. After the appropriate incubation time, arrest the reaction with the addition of 150 μl of 0.2 *M* Na_2CO_3, pH 11.2. In addition to stopping the reaction, the high pH of the sodium carbonate serves to enhance the fluorescence of the released 4-MeU.
7. Read the fluorescence using an excitation wavelength of 355 nm and emission wavelength at 460 nm. The Fluoroskan can be blanked on a single well, an entire column, or a row of wells. Alternatively, the blank values can be subtracted after the plate has been read in the NO BLANK mode.

Figure 3 shows the fluorescence intensity as a function of enzyme concentration at incubation times of 10, 30, and 60 min. An essentially linear response is observed up to 2 ng of GUS. From the slope of each line, which is indicated in Table 1, an average specific activity of 119 fluorescence units/ng/min is obtained. This number is based on the fluorescence values obtained with the Fluoroskan II and will be quite different when measured on the LS 50 plate reader. It can be seen that an incubation time of 30 min at 37°C is quite adequate.

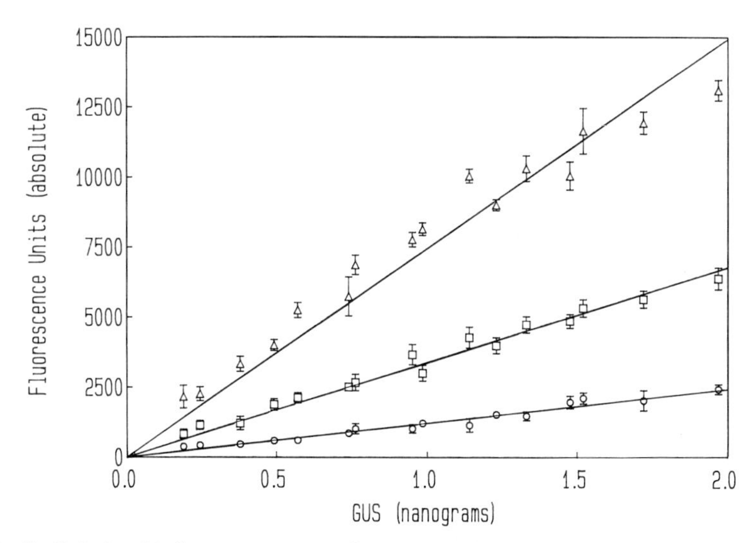

Fig. 3 Relationship between amount of enzyme and observed fluorescence for a fixed time of incubation with substrate (MUG). ○, 10 min; □, 30 min; △, 60 min. See text for details. Each point represents an average value of an experiment peformed in triplicate. Standard deviations are also indicated (from Rao and Flynn, 1990).

Table 1

Specific Activity of GUS as Computed from Data Shown in Fig. 3[a]

Incubation time (min)	Slope[b]	Specific activity[c]
10	1215.58 ± 63.42	121.50
30	3388.87 ± 120.24	112.96
60	7465.10 ± 422.72	124.42

[a] The slope of each curve is divided by the appropriate time of incubation in minutes (from Rao and Flynn, 1990).
[b] Fluorescence units/ng GUS.
[c] Units/ng/min.

In a similar fashion, one can determine the specific activity of GUS when the substrate REG (Molecular Probes, Eugene, Ore.) is used, with some minor modifications:

1. The lysis buffer does not contain β-mercaptoethanol (see step 1 above).
2. The stock solution is prepared as a 1 m*M* solution in 0.1 *M* sodium phosphate buffer, pH 6.5, and only 2.5 μl of it is used per 50-μl assay (see step 4 above).
3. The reaction is stopped (see step 6 above) with 100 m*M* sodium phosphate buffer, pH 7.0.
4. The fluorescence intensity of the product resorufin is measured on the Fluoroskan using an excitation of 544 nm and emission of 590 nm.

When TUG (Molecular Probes, Eugene, Ore.) is used as a substrate, the stock solution is prepared and dispensed as with MUG, but with the following modifications:

1. The reaction is stopped (see step 6 above) with 100 m*M* Tris-HCl, pH 9.0.
2. The fluorescence intensity of the product 4-trifluoromethylumbelliferone is most accurately measured on the LS 50 plate reader using an excitation of 400 nm and an emission of 510 nm. These settings are not available on the Fluoroskan.

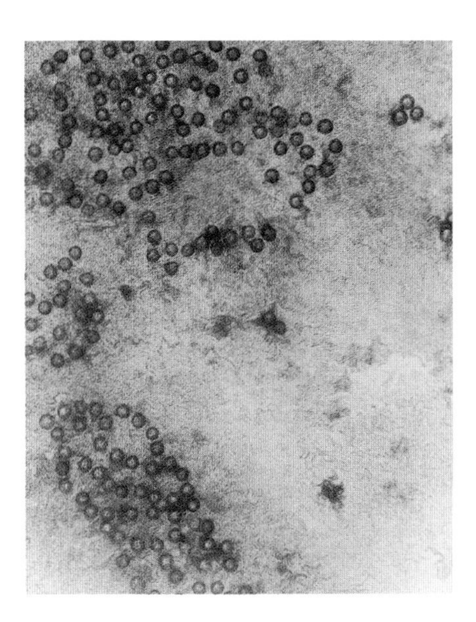

Color Plate 1 Pollen grains of a nontransformed tobacco plant engulfed by a Myxobacteria-like organism on an LB plate containing X-Gluc. (See Chapter 1.)

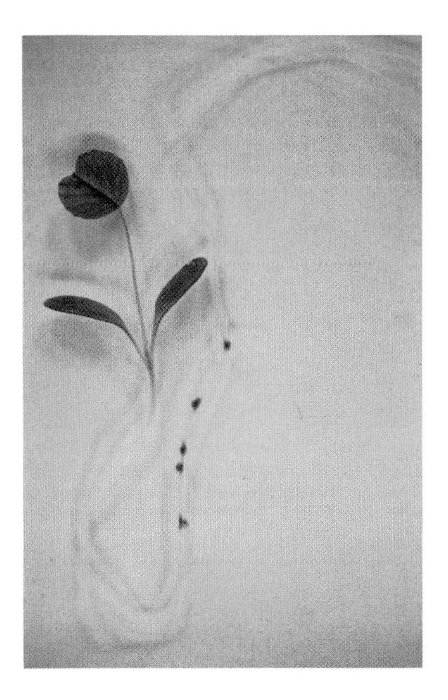

Color Plate 2 Nodules induced on an alfalfa plant by a *R. meliloti* strain carrying the *E. coli gus* operon. The whole root has been incubated in buffer containing X-Gluc, and shows intense blue staining in the nodules caused by the GUS activity of the bacterial occupants. (See Chapter 1.)

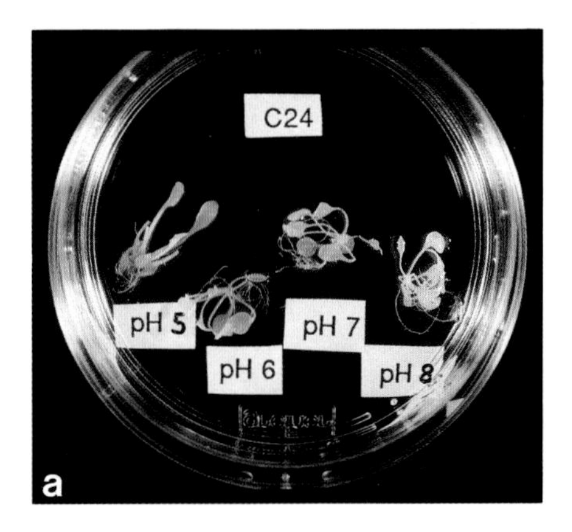

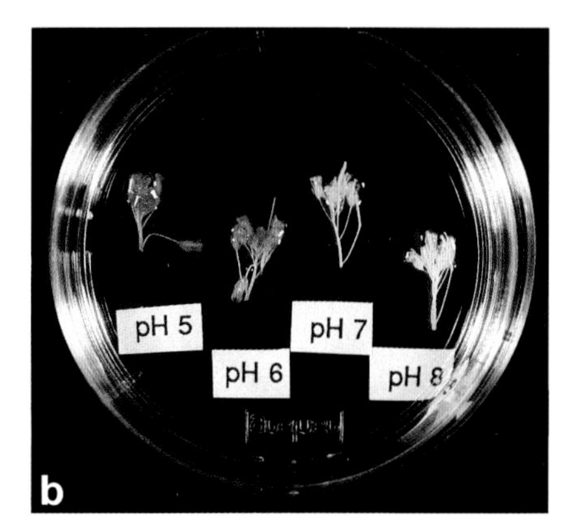

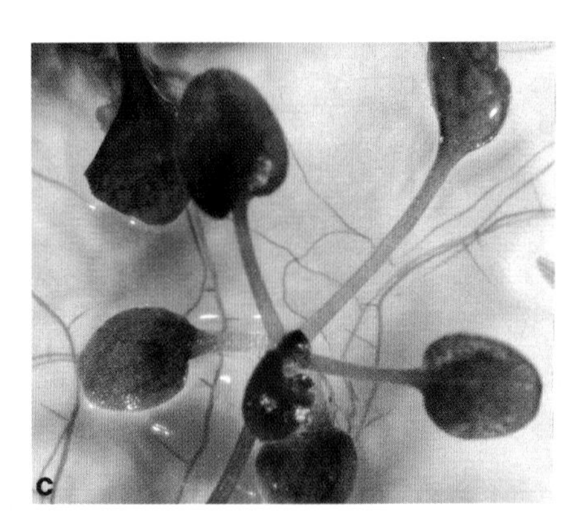

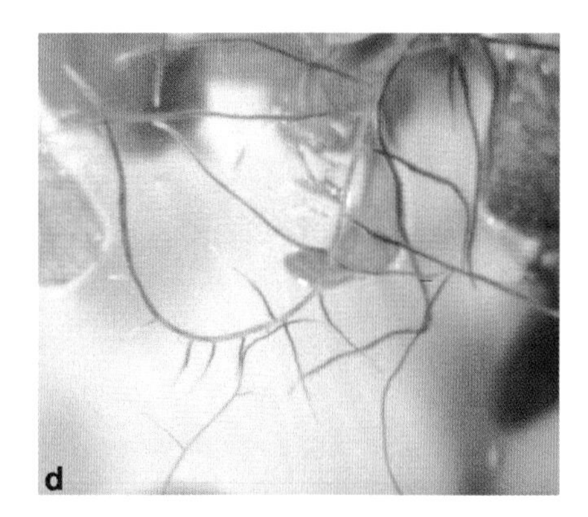

Color Plate 3 X-gluc staining of *Arabidopsis thaliana*. (a, b) Standard protocol for histochemical GUS assay at different pH values; (a) whole wildtype seedlings (C24) stained at pH 5, pH 6, pH 7, and pH 8; (b) flowers of wildtype seedlings (C24) stained at pH 5, pH 6, pH 7, and pH 8. (c, d) Nondestructive X-gluc staining of B33-GUS transformants grown in 3MS medium; (c) view from top; (d) view from below. (See Chapter 2.)

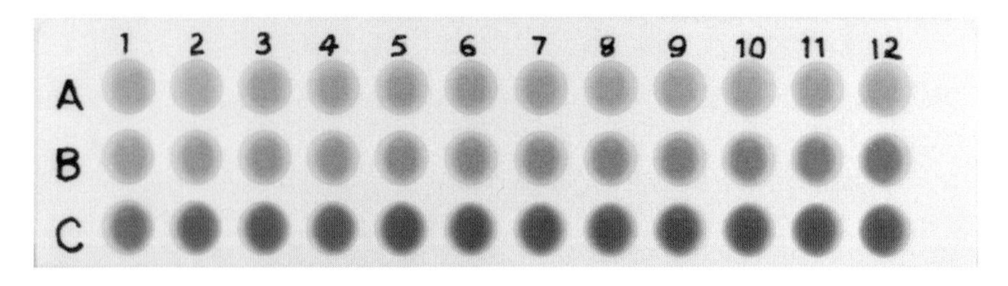

Color Plate 4 The visual detection of GUS activity with REG as substrate. Row A, buffer blank with substrate but no enzyme; rows B and C, with increasing amounts of GUS in increments of 0.02 ng. From left to right, 0.02 ng to 0.24 ng in row B and 0.26 ng to 0.48 ng in row C. (See Chapter 6.)

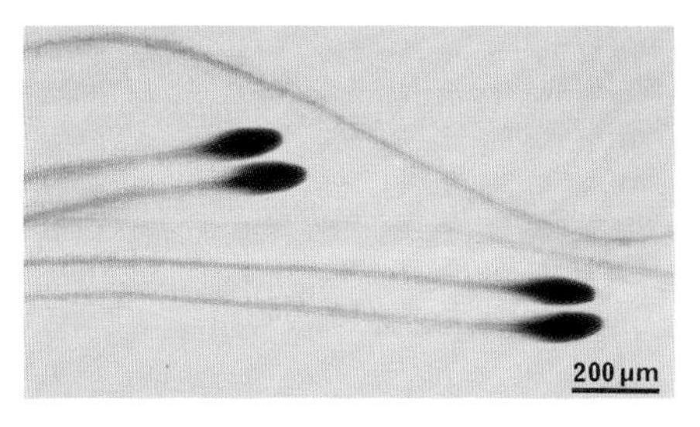

Color Plate 5 Micrograph of root tip tissue from transgenic tobacco expressing a 35S GUS construct. Whole tissue, showing strong expression in the root tip, and weaker expression along the vascular tissue. (See Chapter 8.)

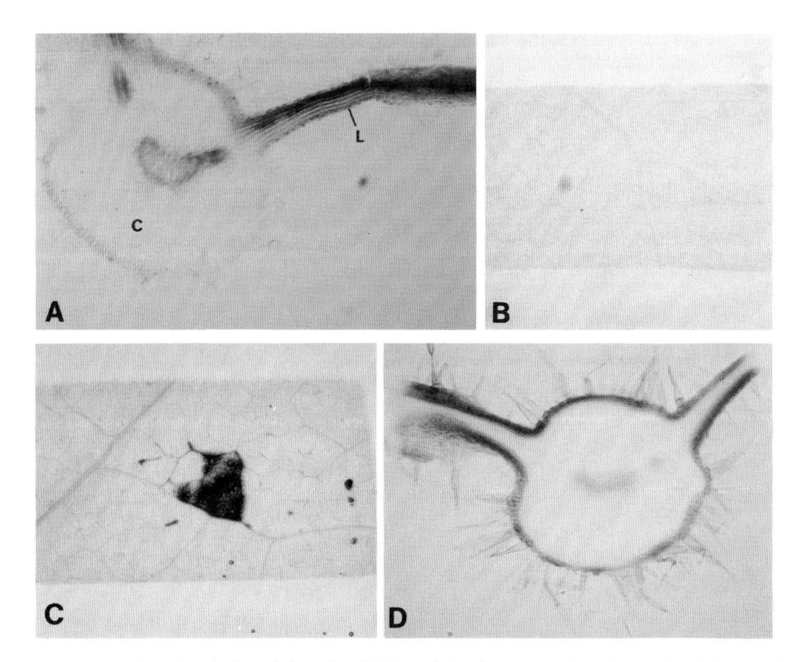

Color Plate 6 Histochemical staining for GUS activity in transgenic tobacco leaf tissue. Tissue sections were cut, fixed, and stained as described (see Protocols) and photographed under a Wild M8 stereomicroscope with a Wild MPS46 camera attachment. The constructs pBI121, pΔ*Ac* G and p*Ac* G are described in Figure 11.3. (A) Transverse section taken from a plant transformed with pBI121. The section was made through the petiole mid-rib region of the leaf base with a short piece of adjacent lamina (L) and shows GUS activity in all tissue except the petiole cortex (C). (B) Surface view of leaf lamina taken from a plant transformed with pΔ*Ac* G. This is a defective *Ac* element which cannot excise (Finnegan *et al.*, 1989). No GUS activity is seen in any tissue indicating that cloning of a transposable element into the untranslated leader of the 35S-GUS-nos3′ chimaera does inactivate the GUS gene. (C) Surface view of a leaf lamina taken from a plant (p*Ac* G-1) transformed with p*Ac* G. Small sectors of localized GUS activity indicate that *Ac* excision occurred late in development. (D) Transverse section (as described in A) of a leaf taken from a plant (p*Ac* G-3) transformed with p*Ac* G. GUS activity is restricted to epidermal cells and attached trichomes, indicating that an excision event occurred early in development in the tunica 1 cell layer of the somatic embryo. (See Chapter 11.)

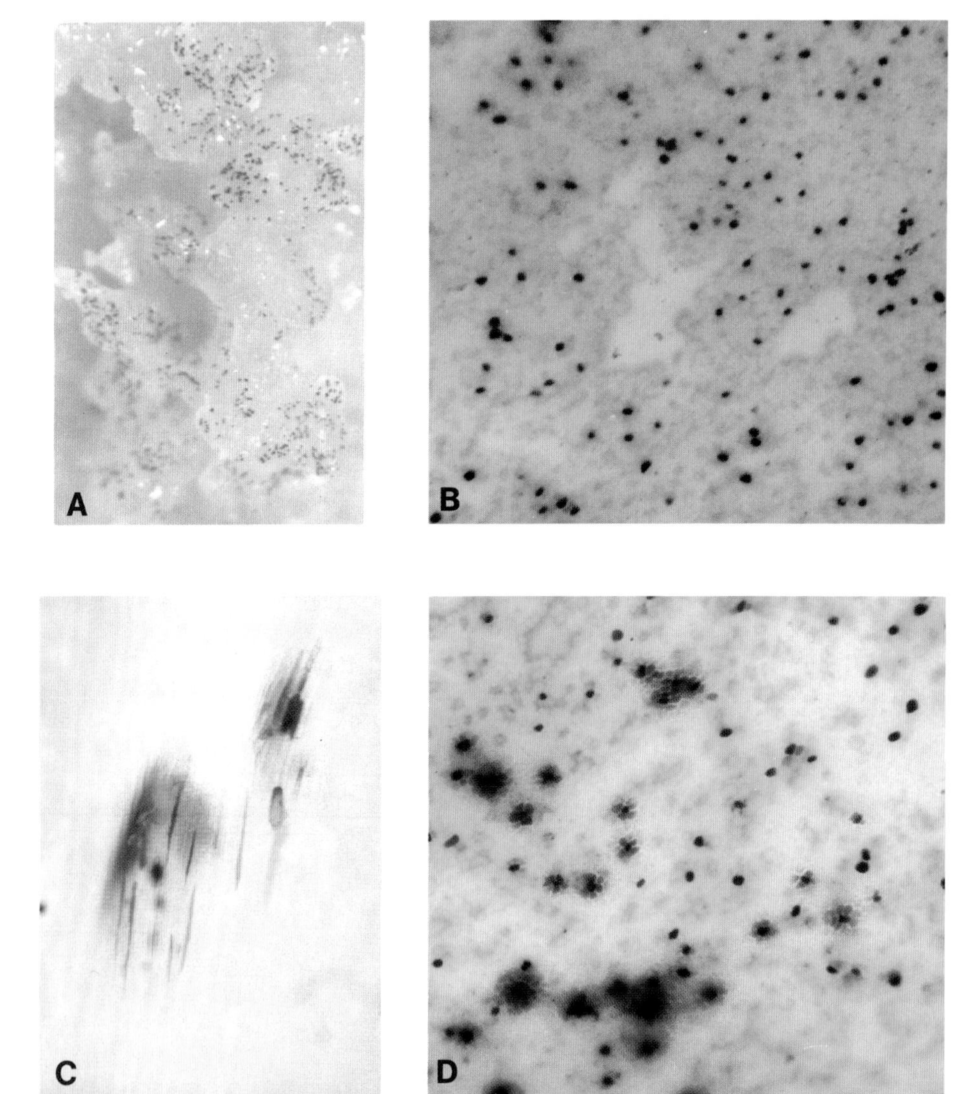
A
B
C
D

Color Plate 7 (A) Anthocyanin accumulation in embryogenic maize suspension cells (c r-r B-b pl) 48 hr after bombardment with a vector containing both 35S::R and 35S::C (pPHI687; Lynne Sims, unpublished). (B) Red cells in the aleurone layer of a maize kernel (C r-g) 48 hr after bombardment with 35S::R (pPHI443; Ludwig *et al.*, 1990). (C) Coleoptile tissue (r-g B-b pl) stained with X-Gluc 72 hr after bombardment with 35S::R (pPHI443) and 35S::GUS (pPHI459; Ludwig *et al.*, 1990). (D) Aleurone tissue (C r-g) treated as in (C). (E) Sector in the fourth leaf of a seedling which developed from an immature maize embryo (C r-g B-b pl) bombarded with 35S::R (pPHI443) 8–10 days after pollination. (See Chapter 12.)

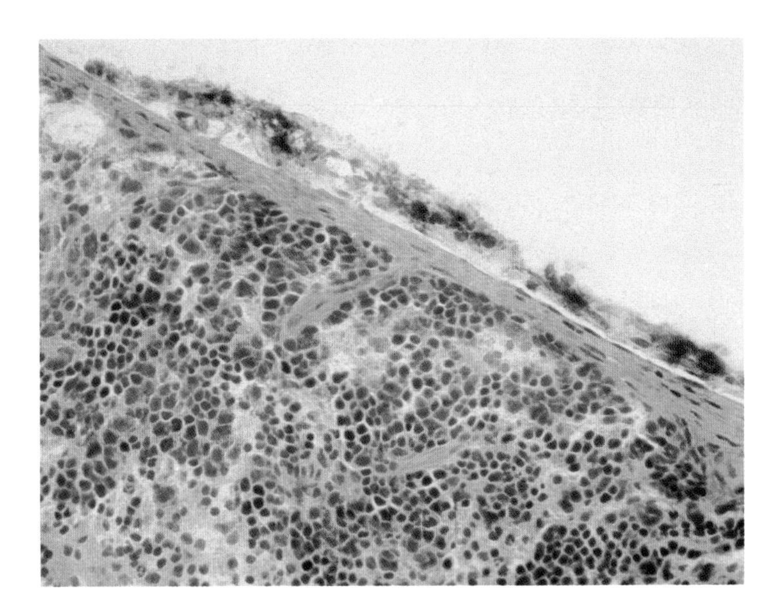

Color Plate 8 Spleen from an MPS VII β-glucuronidase deficient mouse shows β-glucuronidase activity (red) in cells within the splenic capsule 24 hr after intraperitoneal injection of recombinant human β-glucuronidase (Naphthol AS-BI β-D-glucuronide staining with methyl green counterstain, ×100). (See Chapter 14.)

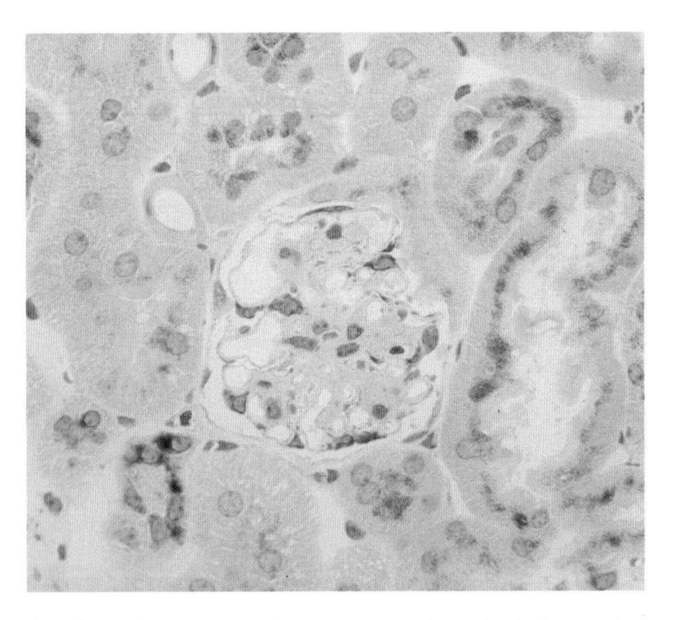

Color Plate 9 Kidney from a transgenic mouse expressing only the human β-glucuronidase transgene shows intense staining for β-glucuronidase (blue) in visceral epithelial cells and the lumenal aspect of renal tubular epithelial cells and weaker staining of mesangial and endothelial cell cytoplasm (X-Gluc staining with Nuclear Red counterstain, ×157). (See Chapter 14.)

Influence of Plant Extracts on GUS Activity

In order to quantitate GUS activity in transformed plant tissue, it is first necessary to determine the optimal amounts of protein to use in the assay. The presence of endogenous inhibitors of GUS activity could lead to an underestimation of GUS if an excessive amount of protein is used. Some degree of background fluorescence may also occur from nonspecific hydrolysis of the substrate (by factors present in the tissue extract) and/or endogenous phenolic compounds.

Background fluorescence, with MUG as substrate, is measured as follows. As an example, untransformed tobacco leaf tissue is used as a source:

1. Punch out a small piece of leaf (~5 mg) in a Kontes tube and homogenize in 50–200 μl of lysis buffer.
2. Pellet cell debris by spinning at 10,000 rpm for 15 min at 4°C.
3. Transfer supernatant to a fresh tube and estimate protein by the method of Bradford (1976).
4. Dispense an aliquot of extract equal to 2–5 μg of protein into the wells of the microtiter plate. Make sure that there are two wells to which no tissue protein is added. These wells will serve as buffer blank.
5. Add lysis buffer to a volume of 45 μl.
6. Initiate the reaction with 5 μl of MUG, incubate at 37°C for 30 min, and stop the reaction with the addition of 150 μl of 0.2 *M* Na_2CO_3.
7. Measure the fluorescence on the Fluoroskan.

Generally, the background fluorescence values average around 5–10 fluorescence units above that of the buffer blank, which is usually around 15 fluorescence units. This appears to hold true for a variety of tissues such as corn embryos, corn callus, corn suspension cells, corn leaf, soybean leaf, sunflower leaf, and tobacco suspension cells.

One can determine the optimal amounts of protein to use by reconstruction experiments in which varying amounts of protein from an untransformed plant are added to the pure enzyme and the activity is measured. The procedure is exemplified below using tobacco leaf tissue as a source of protein.

1–4. Same as above.

5. From the working solution of GUS described earlier, add to each well containing the tissue protein, 0.2–1.0 ng of GUS. Ensure that

there are a few wells which contain tissue protein but to which no enzyme is added, to obtain background fluorescence.

6. Same as above.
7. Measure the fluorescence as before.

Figure 4 illustrates the results of such a reconstruction experiment using 2 and 5 μg of protein from tobacco leaf with MUG (Figure 4A), TUG (Figure 4B), and REG (Figure 4C) as substrates. With the latter two substrates little or no inhibition of GUS activity is observed at both 2- and 5-μg amounts of tobacco protein. However, some inhibition apparently occurs when 5 μg of protein is used with MUG as substrate. We have observed that this inhibitory effect is removed when the tobacco extract is desalted by passing through Sephadex G-25 spin columns (Boehringer-Mannheim) prior to the assay.

GUS Activity in Transformed Tissue

In the determination of GUS activity in transformed tissue, protein amounts causing >10–20% inhibition of pure enzyme activity should be avoided. It is recommended that one not exceed 4 μg of protein.

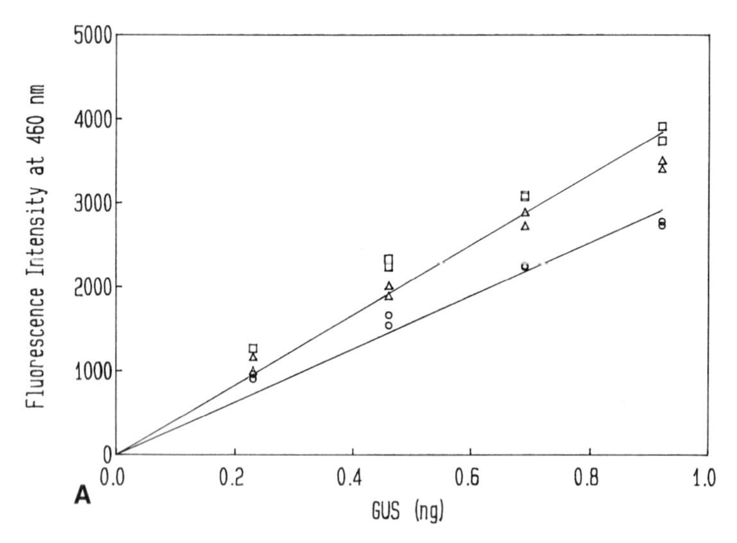

Fig. 4 Effect of tobacco leaf extract on GUS activity using (A) MUG, (B) TUG, and (C) REG as substrates. See text for details. □, Activity of pure enzyme; △, activity in presence of 2 μg of leaf protein; ○, activity in presence of 5 μg of leaf protein.

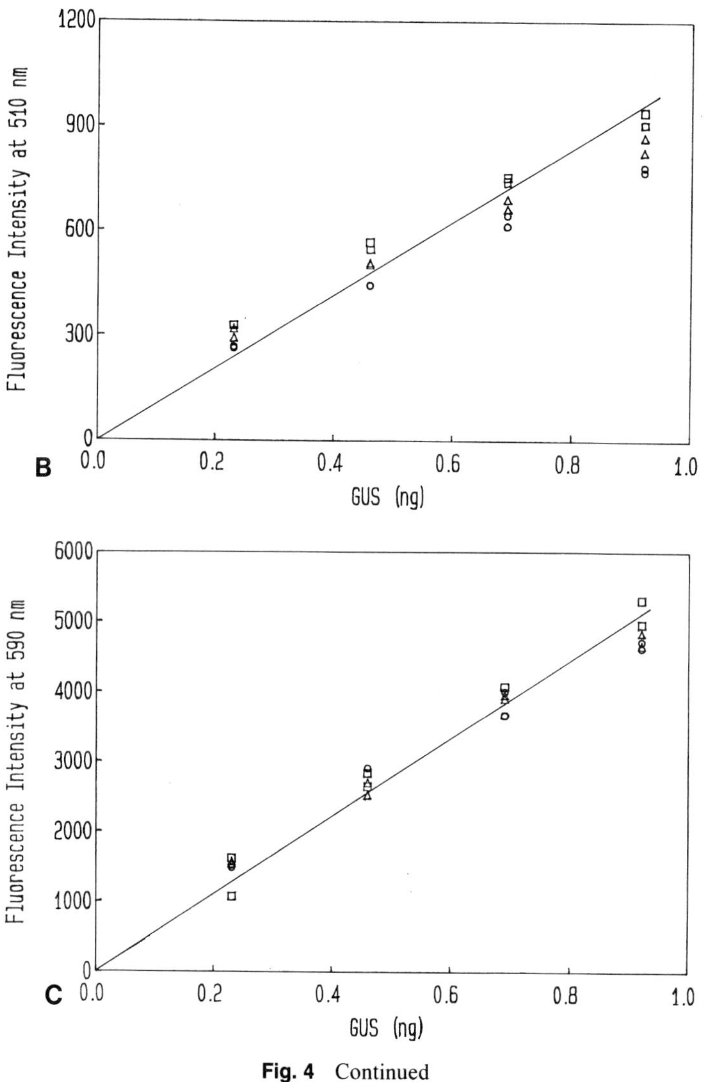

Fig. 4 Continued

1. Homogenize the tissue as described.
2. Determine protein concentration.
3. To the wells add aliquots of tissue extract that fall within the recommended range of protein amount, that is, 4 μg. It is best to use two different amounts of protein. This provides a good cross

check in the final computation of GUS concentration as a function of starting protein.

4. Add lysis buffer, substrate, etc. as described earlier.

It must be noted that only samples showing at least two to four times the background fluorescence can be considered truly positive for GUS activity.

The quantitation of GUS activity is done as follows. If X is the specific activity of the pure enzyme in fluorescence units/ng/min (as determined by the standard curve), Y is the fluorescence of the experimental sample after time T of incubation (in minutes), the amount of GUS is obtained from the equation

$$\text{GUS (nanograms)} = Y/TX$$

For example, if the specific activity is 119 and the fluorescence of the experimental sample is 100 units after 30 min of incubation, the amount of GUS in the sample is $100/30 \times 119$ or 0.028 ng. Further, if 2 μg of total protein were used in the assay the percentage amount of GUS would be 0.0014% or 14 ppm total protein.

Discussion

In the past, determination of GUS activity in transgenic plants has been restricted to the analysis of only one sample at a time owing to the limitation of the instrument itself to measure fluorescence of samples in individual cuvettes or tubes. Therefore, despite the availability of a large variety of spectrofluorimeters in the market (Perkin-Elmer, SLM, Shimadzu, Hitachi, etc.) the intrinsic sensitivity of the fluorimetric assay (Mead *et al.*, 1955; Jefferson, 1987) was offset by the inability to handle more than one sample at a time. However, the recent introduction of the fluorescence microtiter plate reader Fluoroskan II and the model LS 50 spectrofluorimeter with plate reader attachment now permits the simultaneous analysis of many samples. This development, together with the adaptation of the GUS assay to the 96-well microtiter plate (Rao and Flynn, 1990), should facilitate the integration of the entire assay into a robotics system. Such a system would make the processing of a large number of samples less labor intensive, especially since the rate-limiting step in the assay procedure is the tissue grinding (see Brumback, Chapter 5).

Additionally, the flexibility allowed by the monochromator-based LS 50, in choosing any excitation or emission wavelength, will stimulate the design and synthesis of newer substrates. Particularly interesting will be those substrates in which the products have a long excitation wavelength (lower energy) and therefore lower background fluorescence. Also, detection of fluorescence is enhanced when there is a large difference (Stoke's shift) in the excitation and emission maxima of the fluorescent product. Among the substrates currently available, TUG is the only one in which the product, 4-trifluoromethylumbellifrone, has a Stoke's shift >100 nm. Indeed, it has been possible to detect as low as 0.02 ng of the pure enzyme using this substrate on the LS 50.

Perhaps an ideal substrate will be one that will be sensitive enough to allow visual detection of GUS activity, which would make it especially useful for rapid screening of transgenic plants in the field. A substrate that approaches these requirements is REG. In experiments performed with the pure enzyme (Color Plate 4) one could easily detect 0.2 ng of GUS from the disinct pinkish color of the solution. The color became more intense at higher concentrations of GUS. Reconstruction experiments using 5 μg protein from tobacco leaf extracts indicated that 0.5 ng of GUS could be easily detected.

References

Bradford, M. M.(1976). A rapid and sensitive method for the quantification of microgram amounts of protein using the principles of protein-dye binding. *Anal.Biochem.* 72, 248–254.

Jefferson, R. A. (1987). Assaying chimeric genes in plants: The GUS gene-fusion system. *Plant Mol. Biol. Rep.* 5, 387–405.

Jefferson, R. A., Burgess, S. M., and Hirsh, D. (1986). β-Glucuronidase from *Escherichia coli* as a gene-fusion marker. *Proc. Natl. Acad. Sci. USA* 83, 8447–8551.

Mead, J. A. R., Smith, J. N., and Williams, R. T. (1955). Studies in detoxication. 67. The biosynthesis of the glucuronides of umbelliferone and 4-methylumbelliferone and their use in fluorimetric determination of β-glucuronidase. *Biochem. J.* 61, 569–574.

Rao, A. G., and Flynn, P. (1990). A quantitative assay for β-D-glucuronidase (GUS) using microtiter plates. *BioTechniques* 8, 38–40.

PART 3
Histochemical Detection of GUS

7 Histochemical Localization of β-Glucuronidase

Anne-Marie Stomp
Forestry Department
North Carolina State University
Raleigh, North Carolina

Introduction

Histochemical staining of β-glucuronidase was first reported in the early 1950s for localization of endogenous enzyme in mammalian tissue (Fishman, 1955; Conchie *et al.*, 1959). The preferred substrate for localization is 5-bromo-4-chloro-3-indolyl-β-D-glucuronide, or X-Gluc. Pearson *et al.* (1961) developed this colorless substrate because of the high extinction coefficient (making it readily detectable at low concentrations) and aqueous insolubility of the final cleavage product, dichloro-dibromoindigo (ClBr-indigo). The reaction proceeds through an unstable intermediate, which then undergoes an oxidative dimerization to the intensely blue ClBr-indigo (Figure 1). This second characteristic makes X-Gluc ideal for localization because ClBr-indigo immediately precipitates upon formation, allowing precise cellular localization of enzymatic activity and little loss of enzyme product, ClBr-indigo, in solvents typically used during tissue processing.

β-Glucuronidase activity, although reported in *Scutellaria baicalensis* and almonds (Fishman, 1955), is not considered a normal part of the enzyme complement of higher plants, and has historically held little interest for plant biologists. However, cloning of the β-glucuronidase gene (gus A) from *E. coli* and construction of chimeric genes and plasmid vectors for plant transformation (Jefferson *et al.*, 1986, 1987) have dramatically increased interest in histochemical localization methods for

GUS Protocols: Using the GUS Gene as a Reporter of Gene Expression

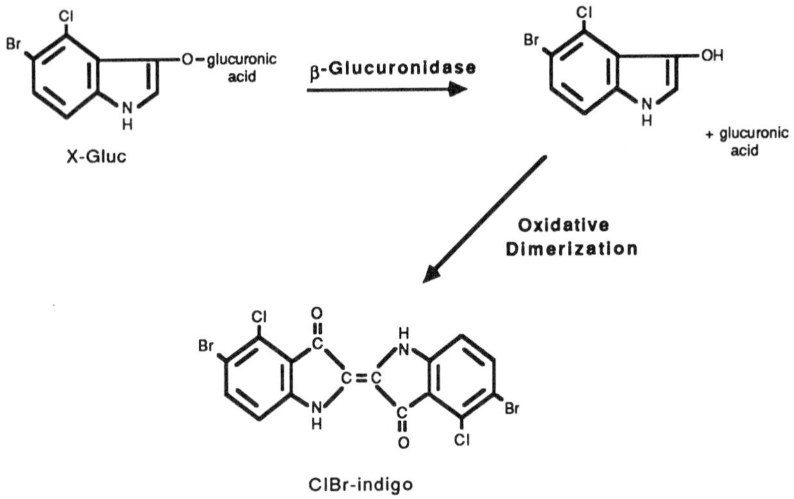

Fig. 1 Clevage of 5-bromo-4-chloro-3-indolyl-β-D-glucuronide, or X-Gluc, produces the final insoluble blue precipitate dichloro-dibromoindigo (ClBr-indigo).

this enzyme. GUS is a useful marker in plants because under the proper assay conditions histochemical staining gives low or no background and is quite a forgiving and easy assay to perform.

The purpose of this chapter is to outline basic protocols for histochemical staining of GUS using X-Gluc, including tissue preparation, staining reagents, sectioning, and photomicrography. Following the basic protocol, I will discuss the latitude allowed in this assay and some of the problems that we have encountered using this procedure in our own research on trees.

Protocols

Tissue Preparation

Figure 2 gives a flow chart showing the decisions that need to be made for histochemical staining. First, one needs to decide whether to fix the tissue prior to staining. Unless the particular gene construct targets the enzyme to a subcellular compartment, GUS enzyme is localized in the cytoplasm of transgenic plant cells (Iturriaga *et al.,* 1989). Staining of unfixed thin sections, such as cryostat sections, can result in poor

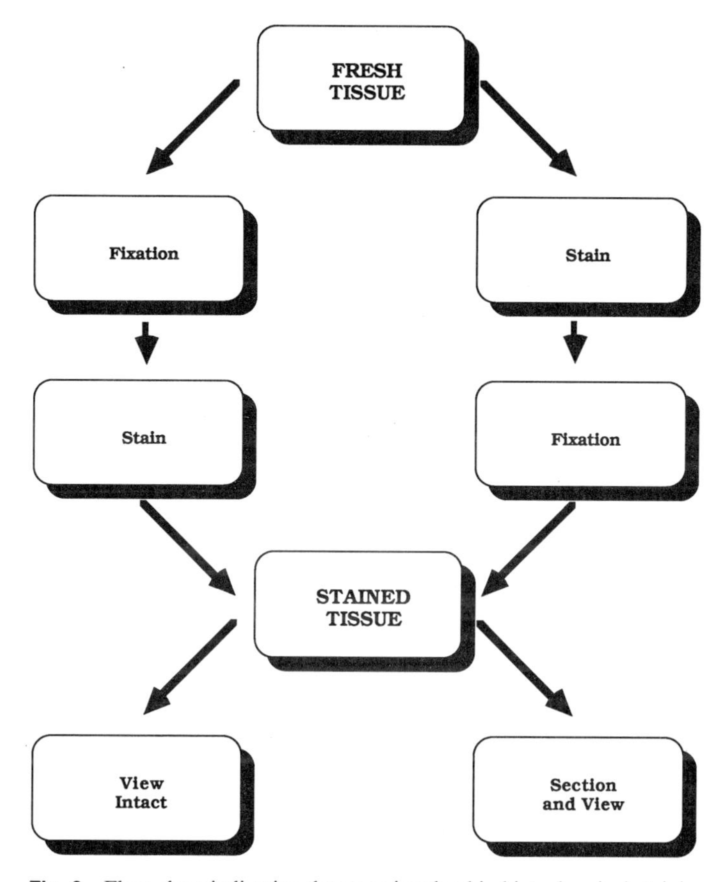

Fig. 2 Flow chart indicating the steps involved in histochemical staining.

localization due to leakage of the enzyme from cut cells. This problem is inversely proportional to the section thickness, as thiner sections (e.g., 10–20 μm) will be composed almost entirely of cut cells. Therefore, if staining unfixed sections, cut sections that are at least three times thicker than the average cell diameter, for example, hand sections or Vibratome sections.

If unfixed tissue is to be stained, it is highly desirable to stain intact tissue or large blocks of tissue to minimize loss of enzyme. However, differentiated plant organs, such as leaves and stems, can be covered with a heavy cuticle, which will slow penetration of staining reagent mix. Therefore, intact tissue staining is best done with very young (e.g., embryos, newly expanding leaves), succulent material. Older material should be cut into pieces of not greater than 1 cm. Staining will occur in pieces larger than this but the results may not truly reflect the level of GUS activity present deep within the tissue.

Tissue Fixation

Tissue can be fixed before or after staining (see also Martin *et al.*, Chapter 2, and Craig, Chapter 8).

To fix tissue before staining, place in a buffered 2–3% gluteraldehyde solution on ice for 2–3 min, or in a buffered 4–6% formaldehyde solution on ice for 2–5 min.

Short fixation times are important so as not to loose the activity of the glucuronidase. It is important to use histological-grade aldehyde solutions, available in several percent concentrations, sealed under inert gas, or use solid paraformaldehyde. These ampules should be stored refrigerated, never frozen, until used. The buffer of choice for plant material is 0.05 *M* potassium phosphate buffer, pH 7.0; however, sodium phosphate buffer is commonly used at concentrations varying from 0.05 to 0.1 *M*.

Fixing solution is commonly made from a 2× stock solution of gluteraldehyde in distilled water. To make the 2× stock, place the broken vial into the appropriate amount of distilled water, because concentrated gluteraldehyde solutions are thick. Occassionally, the working stock will turn milky as the gluteraldehyde goes into solution, but it should clear upon standing. Never use milky or discolored gluteraldehyde. The working stock solution can be stored refrigerated for 1–2 months. Long-term storage of diluted gluteraldehyde solutions is not desirable, due to oxidation of the aldehyde. Fixing solution should be freshly prepared by diluting the 2× working stock 1 : 1 with 2× buffer. Fixation after staining can be done using any of the common procedures for plant material, such as FAA, gluteraldehyde, or ethanol.

A convenient way to fix and dehydrate stained tissue is by freeze substitution with ethanol, essentially following the protocol developed by Meyerowitz (1987) for *in situ* hybridization.

1. Place previously stained tissue in 100% ethanol sitting in a dry ice, 100% ethanol bath.
2. Transfer the ethanol immersed tissue to a −70°C freezer for 24–48 h (depending on tissue size) to allow the exchange of ethanol for water in the tissue.
3. Slowly bring the tissue to room temperature, with intervening incubations of several hours duration at −20°C and 4°C.

4. Place the tissue into fresh 100% ethanol.
5. Incubate at room temperature for several hours, and then give two 2-h passages in toluene.
6. Transfer the tissue to molten paraffin (Tissueprep or Paraplast) and give two passages of several hours each.
7. Embed the tissue for sectioning.

It is important that the tissue sink in toluene before moving into wax, indicating complete exchange of ethanol for water and toluene for ethanol. The times and number of passages in each solvent need to be determined for each tissue and will vary due to tissue size, surface area, and the degree to which the surface area is covered by cuticle. Plastic embedding can also be done with GUS-stained material. We have embedded prestained material in either Spurrs (using only ethanol solutions for dehydration) or LX-112 (using ethanol solutions for dehydration and propylene oxide for transitional solvent) and have had no appreciable loss of stain. GUS staining is clearly visible in tissue stained with osmium tetroxide in preparation for transmission electron microscopy (TEM). This is useful for locating GUS-expressing cells on block faces prior to final trimming for ultrathin sections needed for TEM studies of transgenic cells.

Staining

The reagent mix typically contains three components: the substrate (1–5 m*M* X-Gluc), the buffer (0.1 *M* sodium phosphate, pH 7.0), and the oxidation catalyst (0.5 m*M* each potassium ferri- and ferrocyanide, pH 7.0, plus 10 m*M* EDTA, pH 7.0) (Table 1).

Table 1

Typical Reagent Mix

Stock solution	Final concentration	Reagent mix (μl/ml)
1.0 *M* $NaPO_4$ buffer, pH 7.0	0.1 *M*	100
0.25 *M* EDTA, pH 7.0	10 m*M*	40
0.005 *M* KFerricyanide, pH 7.0	0.5 m*M*	100
0.005 *M* KFerrocyanide, pH 7.0	0.5 m*M*	100
0.02 *M* X-Glucuronide	1.0 m*M*	50
10% Triton X-100 (optional)	0.1%	10
Subtotal		400
Distilled water		600
Final volume		1000

Substrate

The substrate, 5-bromo-4-chloro-3-indolyl-β-D-glucuronic acid, or X-Gluc, is the most expensive component of the reagent mix. We have found no difference in the quality of X-Gluc available from several suppliers. X-Gluc is supplied as the cyclohexylammonium salt and is stored frozen. Stock solution is made up as 0.02 *M*, most typically in *N,N′*-dimethylformamide; dimethyl sulfoxide (DMSO) will also work. In our hands, DMSO solutions are somewhat less stable than those made in dimethylformamide, but certainly quite adequate. Stock solutions are aliquoted into convenient amounts (for a final concentration of 2.0 m*M* we find 100-μl aliquots useful) and stored in 0.5-ml microfuge tubes at −20°C. We have stored stock solutions for up to 6 months with no problems. Degraded stock solutions are recognized by a brown to purple color change. Certain plastics, such as polystyrene, will react with dimethylformamide and should not be used.

Oxidation Catalyst

The potassium ferricyanide and ferrocyanide are present in the reaction mix to accelerate the oxidative dimerization of the colorless cleavage intermediate into the colored final product, ClBr-indigo (Lojda, 1970b). Freshly prepared ferricyanide will be strongly yellow and ferrocyanide will be a very pale yellow. Upon standing, ferrocyanide will turn darker yellow due to oxidation of the ferrous ion to the ferric ion. Although the partially oxidized ferricyanide solution can be used, we do not use solutions more than 2 months old. The ethylenediamine tetraacetic acid (EDTA) is added to mitigate the partial inhibition of the enzyme by the oxidation catalyst.

Reagent Mix

There are a variety of ways the reagent mix can be prepared and stored. We keep separate stock solutions of each component, storing all but the X-Gluc in the refrigerator. We always prepare fresh reagent mix, determining the final volume required and from that calculating the amounts of the various components and assembling the mix. The advantage is that if a stock solution goes bad we do not have to prepare new stock solutions for all the components. Alternatively, a single large batch of reagent mix can be made in advance and frozen (for at least 6 months at −20°C) in convenient aliquots. This minimizes mixing errors, but can be expensive if an error is made in mixing or the large batch gets contaminated, necessitating disposal of the mix with the X-Gluc already added. A third possibility is to mix all components except the X-Gluc, bringing it to 95% volume. A 20× X-Gluc stock is then added just before use. We know people who have used all these approaches successfully.

Incubation Conditions

1. Place tissue to be stained, either fixed or unfixed, in enough staining reagent mix to completely cover the tissue. Microtiter dishes are quite convenient as staining vessels. If tissue wetting is a problem, Triton X-100 can be added to a final concentration of 0.1%.
2. Cover the tissue to prevent evaporation of the reagent mix and then incubate at 37°C. We wrap the microtiter plates with plastic wrap, which works better than parafilm because it will not crack at 37°C.
3. Incubation times vary greatly. Blue color can start to appear in as little as 10 minutes with certain promoters and may take several hours to develop with others.

The assay does not suffer from overly long incubations (e.g., 24 h), however, plant tissue will often turn brown with long incubation times. This makes faint blue cells or tissue spots much more difficult to see. For best results, the minimal incubation times should be used.

Staining Considerations

The only necessary component to achieve GUS staining is the substrate, diluted in water. The problem in using only the substrate is that diffusion of the undimerized intermediate can occur (especially if incubation conditions are largely anaerobic), leading to an overestimate of the transgenic cell population size. Leaving out the buffer or using a concentration of buffer with insufficient buffering capacity has resulted in readily detectable levels of background "endogenous" GUS activity in our hands, both with Norway spruce embryos and loblolly pine cambial tissue.

Background GUS activity has also been reported in a wide variety of plants and plant parts (Hu *et al.*, 1990), using 50 m*M* sodium phosphate buffer. Lojda (1970a) showed that slow cleavage of 5-bromo-4-chloro-3-

indolyl-β-D-glucoside (a compound closely related to X-Gluc) occurs at mildly acidic pH (5.5–6.0), and that staining due to this cleavage can be intensified by the presence of peroxidases. Peroxidases are ubiquitous in plant tissue and unbuffered or weakly buffered, sectioned or damaged plant tissue could readily have a pH between 5.5 and 6.0 due to leakage from cell vacuoles. It is interesting that Hu *et al.* (1990) report profiles of endogenous GUS activity in germinating seeds of both soybean and stringbean, which peak at 4 days. Peroxidase activity in germinating seedlings is known to peak between 3 and 6 days (Doby, 1965). We have estimated the size of the "endogenous" GUS activity in Norway spruce to be roughly 60,000 molecular weight, well within the range of some plant peroxidases (Kim *et al.,* 1988). Endogenous GUS activity in Norway spruce and loblolly pine cambium is minimized or completely abolished if the substrate solution is buffered with 0.1 M sodium phosphate buffer, pH 7.0, and/or if the tissue is heated for 5–10 min at 60°C. Therefore, staining should be done in no less than 0.1 *M* buffer at pH 7.0, and if "endogenous" activity is a problem, investigate the potential of pH manipulation to reduce background, artifactual staining.

Lojda (1970b) developed the ferri-/ferrocyanide oxidative catalyst for this histochemical assay. In the mammalian tissue examined, the catalyst abolished low-level staining background and gave more limited enzyme localization, which did not mimic localization of tissue peroxidase. Staining can be done without the oxidative catalyst, including the absence of EDTA; however, localization will not be as precise.

An important disadvantage of the complete GUS staining procedure is that it kills tissue if performed as written. We have investigated the toxicity of the staining components and the incubation conditions to develop a nondestructive assay for pine cotyledons on shoot inductive medium. We have been moderately successful in using only substrate in water or tissue culture medium, for minimal incubations (1–2 h) at 37°C. The most toxic component in the reagent mix is the 0.1 *M* sodium phosphate buffer. Substituting potassium for sodium increases the survival time of the tissue, as does decreasing buffer strength (down to 20 m*M;* the problems with this have already been discussed). DMSO or *N,N'*-dimethylforamide up to 15% (v/v), and the oxidative catalyst mix were not toxic to pine cotyledons. Incubation at 37°C was also quite damaging to the tissue. Incubation at room temperature was much less damaging but resulted in very poor staining; therefore, it is not recommended.

Most staining is done under nonsterile conditions. Therefore it is important to remember that some bacteria and fungi have β-glucuronidase activity. Most importantly for plant transformation work, certain

CaMV35s promoter constructs give low levels of expression in *Agrobacterium* strains. Therefore, early staining to confirm transformation can be misleading if the tissue is not completely free of contaminating bacteria, which usually takes weeks of culture on medium containing high levels (500 mg/l) of carbenicillin and/or cefotaxime. Obviously, other methods of transformation, such as electroporation or microprojectile bombardment, are not affected by this problem.

Data Analysis and Photomicrography

Histochemical GUS staining is used for a variety of purposes in both transient expression systems and with stably transformed cell populations. When optimizing DNA transfer protocols, quantifying the frequency of GUS expression centers or GUS-positive "blue spots" is commonly done, either early after DNA transfer looking at transient expression, or later in a protocol if looking for stably transformed cell populations. Histochemical staining is also a useful tool for promoter studies designed to identify DNA sequences that are involved in environmental, physiological, or tissue-specific expression. In these studies, stably transformed cells, tissues, or plants are used and qualitative data is collected of GUS expression in response to environmental signals, developmental stage, or cell location within a tissue. Visual data is collected and recorded as photographs of whole material or microscopic sections.

Reproducibility in quantifying GUS staining data is determined by the accuracy of detecting blue stained cells within the somewhat opaque plant tissue and above background color, either green or brown (polymerized phenolics). A tradeoff exists between shortening the staining incubation time, to minimize phenolic polymerization, and lengthening the incubation time, to insure detection of low-level GUS expression. Fresh material that has only been stained before viewing is often dark green or brown, and faint blue stain will be missed. Clearing the tissue by incubation in 100% ethanol will remove chlorophyll without removing ClBr-indigo; however, the browning due to oxidative polymerization of phenolics is permanent.

Other problems that make detection of GUS expression difficult are endogenous X-Gluc degradation in some tissues and apparent slow inactivation of the gene in certain species. Endogenous X-Gluc degradation will make detection of low levels of GUS expression difficult if not impossible. We have been sucessful in reducing this activity to a minimum by using 100 m*M* buffer (for staining of loblolly pine xylem and Norway spruce embryos) and in briefly heating tissue to 60°C for 15 min (Norway spruce embryos). Apparent inactivation of the GUS gene is more difficult. We have seen no diminishing of GUS expression in transgenic sweetgum cultures, either undifferentiated nodule cultures or shoots regenerated from these cultures. In contrast, we have seen a marked decrease in the intensity of GUS staining in transgenic loblolly pine cultures after several months in culture. Therefore, when working with a new plant species it is important to remember that GUS staining may not be a reliable marker for determining stable transformation. Quantitative assay using the fluorometric method is much more sensitive for detecting low levels of GUS expression and should be used in conjunction with histochemical staining for determining stable transformation.

Photomicrography of GUS-stained material is straightforward and can yield informative and beautiful results in both black and white and color if several factors are kept in mind. When using the stereomicroscope, we have found that if the tissue is submerged under a thin film of water (reduces glare), if the light angle is very low (reduces glare and somewhat backlights the tissue), and if the phototube diaphragm is closed down (to increase depth of field) we get the best photographs.

Photographing sectioned material under the higher magnification of the compound microscope is somewhat more difficult, but still quite accessible. The most important decision to be made is whether to stain the sections before viewing. This will be determined by the optical systems that are available for use. If only bright field is available, the sections will need to be stained in order to produce enough contrast to visualize cell structure of the tissue. Safranin-fast green is a standard plant cell stain (Johansen, 1940) and works well. Obviously, blue stains, such as toluidene blue, should be avoided. If phase contrast is available, sections need only be cut and mounted (cut, dewaxed, and mounted if wax embedded) and viewed. However, we have found the best photographs can be taken using unstained material and interference contrast optics with polarized light. This optical system enhances the contrast of the unstained cellular structure without adding color in addition to that of the indigo.

References

Conchie, J., Findlay, J., and Levvy, G. A., (1959). Mammalian glycosidases: Distribution in the body. *Biochem. J.* 71:318–325.

Doby, G., (1965). "Plant Biochemistry." Wiley Interscience, London, p. 618.

Fishman, W. H. (1955). β-Glucosidase. *Adv. Enzymol* 16:361–409.

Hu, C.-Y., Chee, P. P., Chesney, R. H., Zhou, J. H., Miller, P. D., and O'Brien, W. T. (1990). Intrinsic GUS-like activities in seed plants. *Plant Cell Rep.* 9:1–5.

Iturriaga, G., Jefferson, R. A., and Bevan, M. W. (1989). Endoplasmic reticulum targeting and glycosylation of hybrid proteins in transgenic tobacco. *Plant Cell* 1:381–390.

Jefferson, R. A., Burgess, S. M., and Hirsch, D. (1986). β-Glucuronidase from *E. coli* as a gene fusion marker. *Proc. Natl. Acad. Sci. U.S.A.* 83:8447–8451.

Jefferson, R. A., Kavanagh, T. A., and Bevan, M. W. (1987). GUS fusions: β-Glucuronidase as a sensitive and versatile gene fusion marker in higher plants. *EMBO J.* 6:3901–3907.

Johansen, D. A. (1940). "Plant Microtechnique." McGraw-Hill, New York, pp. 80–81.

Kim, S.-H., Terry, M. E., Hoops, P., Dauwalder, M., and Roux, S. J. (1988). Production and characterization of monoclonal antibodies to wall-localized peroxidases from corn seedlings. *Plant Physiol.* 88:1446–1453.

Lojda, Z. (1970a). Indigogenic methods for glycosidases I. An improved method for β-D-glucosidase and its application to localization studies of intestinal and renal enzymes. *Histochemie* 22:347–361.

Lojda, Z. (1970b). Indigogenic methods for glycosidases II. An improved method for β-D-galactosidases and its application to localization studies of the enzumes in the intestine and in other tissues. *Histochemie* 23:266–288.

Meyerowitz, E. M. (1987). Experimental protocols: *In situ* hybridization to RNA in plant tissue. *Plant Mol. Biol. Rep.* 5(1):242–250.

Pearson, B., Andrews, M., Grose, F. (1961). Histochemical demonstration of mammalian glucosidase by means of 3-(5-bromoidolyl)-β-D-glucopyranoside. *Proc. Soc. Exp. Biol.* 108:619–623.

8 The GUS Reporter Gene—Application to Light and Transmission Electron Microscopy

Stuart Craig
CSIRO
Division of Plant Industry
Canberra, Australia

Introduction

The introduction of foreign genetic material into plants is now commonplace. *Agrobacterium tumefaciens, A. rhizogenes,* particle guns, membrane fusion with polyethylene glycol, and electroporation have all been used successfully with tissues ranging from leaf discs, to protoplasts, embryos, and pieces of callus. Until recently, assaying for successful transformation was extremely time-consuming; because of the need for large samples, tissues had to be grown on selective media for weeks or months or a protoplast had to be regenerated into callus or a plant to provide sufficient material for assaying. Radioisotopic methods, for example, the CAT (chloramphenicol acetyl transferase) assay (Hererra-Estrella *et al.,* 1983) provided an excellent procedure for determining successful transformation at the tissue-homogenate level. However, when improved resolution was required—for example, identifying small transformed regions in a larger organ or plant part—the CAT assay proved unsatisfactory since it could not be adapted to histochemical localization.

To optimize the effectiveness of genes in heterologous hosts, it is often necessary that the gene be expressed in the required tissue(s) of the plant, at the required stage of development, and in the required

GUS Protocols: Using the GUS Gene as a Reporter of Gene Expression

quantity. The gene may also need to be activated by stimuli such as insect attack, drought, salinity, etc.

Much of the molecular information that influences the prerequisites listed above resides 5′ to the coding region of the gene in the promoter. Also, it is likely that it will be necessary to construct "designer" genes comprising, for example, the coding region of one gene linked to regulatory sequence(s) from a second gene. Thus, a major current thrust in molecular biology is to gain a detailed understanding of the function of a range of 5′ sequences. To this end, the bacterial enzyme β-glucuronidase (GUS) (Jefferson *et al.*, 1986; Jefferson, 1987) has proven invaluable as a reporter of gene activity. GUS is a hydrolytic enzyme that, when supplied with an appropriate substrate, produces an intense blue, highly insoluble reaction product (Jefferson, 1987).

Detection of chimeric genes containing the GUS reporter gene can be made at the whole tissue level (Color Plate 5), at the cellular level (Figure 1), and at the subcellular level (Figure 2), corresponding to observation with the stereomicroscope, the compound light microscope, and the electron microscope, respectively. Detailed descriptions of our protocols for whole-tissue observations are included in Chapter 11 by Finnegan and by Finnegan *et al.* (1990). In this chapter, use of the technique at both the cellular and subcellular levels is described.

Protocol

Tissue Preparation

Reagents

0.1–1% Glutaraldehyde, freshly prepared from a 70% stock
25 m*M* sodium phosphate buffer, pH 7.0

Tissue preparation involves a tradeoff between retaining enzyme (GUS) activity and preserving tissue structure—the two are generally contradictory. Optimal tissue preservation may result in loss, either partial or total, of enzyme activity. When using a new tissue system, begin with 0.1% glutaraldehyde (GA) and include an unfixed sample to check for loss of enzyme activity due to chemical fixation.

Fixation

1. Under GA in phosphate buffer, cut tissue into pieces using new double-edged razor blades; 1–2 mm^2 is a suitable size to facilitate entry by diffusion of fixative and, later, of substrate. Fix for 30 min following vacuum infiltration of GA at either 22°C or on ice. We have observed no effect of temperature during fixation.
2. Rinse tissue five times in phosphate buffer, 2–3 min per change.

GUS Assay

Stock Reagents

100 m*M* sodium phosphate pH 7.0
100 m*M* potassium ferricyanide
100 m*M* potassium ferrocyanide
5-Bromo-4-chloro-3-indolyl glucuronide (X-glucuronide, Biosynth AG, Stadt, Switzerland) dissolved in dimethyl formamide at 100 mg/ml.

The cyanide solutions should be stored at 4°C in the dark and the X-gluc at −20°C under argon or nitrogen gas. Substrate is prepared by mixing stock reagents to give the following concentrations:

10 m*M* phosphate buffer
0.5 m*M* potassium ferricyanide
0.5 m*M* potassium ferrocyanide
1 m*M* X-glucuronide

Protocol

1. Immediately following the postfixation wash, cover tissues with a minimal volume of substrate mixture and vacuum infiltrate. Store unused substrate solution in the dark at −20°C. A stock of substrate may be stored frozen and aliquots removed for use. We have successfully reused substrate up to three times, but caution is required due to the possibility of bacterial growth.
2. Incubate in the dark at 37°C for 12–24 h; overnight is a convenient period.
3. When the tissue is an intense blue color, terminate the reaction by replacing the substrate with buffer for 5 min.

Jefferson (1987) mentions that GUS may be inhibited by some divalent metal ions and advises the addition of ethylenediamine tetraacetic acid (EDTA) (10 m*M*) to the substrate solution. If large tissues are being assayed without fixation, dimethyl sulfoxide can be added at 1–2% by volume to facilitate entry of the substrate mixture.

Microscopy

Whole Tissue

The GUS-positive tissue can be examined without further processing, or, preferably, following removal of pigments and transfer to an optically favorable medium such as glycerol or immersion oil. The chlorophyll present in most plant organs can totally mask weakly developed sites of GUS activity. We routinely remove pigments from the tissue by passage through 25, 50, 70, 95 and 100% ethanol (10–30 min per step, depending on the size of the pieces, with several changes in 100%). The tissue can be rehydrated through the same series of ethanol and progressively infiltrated with glycerol. A final vacuum infiltration to remove any small residual air bubbles improves the image.

Alternatively, tissue in 100% ethanol can be transferred through acetone to a 1 : 1 mixture of acetone/immersion oil before pure oil. Application of vacuum removes traces of solvent and any small air bubbles. Some loss of reaction product color has been observed when tissue is held in acetone for extended periods.

A third approach has been to embed tissue in casting resin using acetone as the transitional solvent. The tissue can be polymerized in a flat mold or between a microscope slide and coverslip. This approach has the advantage of permanency.

Cellular and Subcellular Observations

For examination of sectioned tissue by light or electron microscopy, great care is required in handling the tissue; tissues should be moved in liquid using a Pasteur pipette and not squeezed with forceps. Also, other forms of shock such as thermal, osmotic, or dessication should be avoided.

Tissue fixation and the histochemical GUS assay are the same as for whole tissue, although one should endeavor to use smaller pieces of tissue to improve the tissue preservation and quality of the sections. For electron microscopy, tissues should be <1 mm^3; slightly larger is acceptable for light microscopy.

1. Following development of the blue color, wash tissue briefly in 10 m*M* phosphate buffer.
2. Refix in 3% GA in 25 m*M* Na phosphate buffer (pH 7.0–7.1) at 22°C for 1–2 h (depending on the size of the tissue pieces).
3. Wash tissue in phosphate buffer (4 × 5 min) then post-fix in 1% osmium tetroxide in the same buffer for 1–2 h at 22°C.
4. After further buffer washes (4 × 5 min), dehydrate tissues as for whole tissue through a graded series of ethanol.
5. Embed in a suitable resin, such as L R White (an acrylic) or Spurr's (an epoxy). Progressively introduce the resin into the tissue via 2 : 1 and 1 : 2 mixtures with ethanol. Infiltration in resin for 1–5 or more days before polymerization may be required to facilitate sectioning. L R White resin is polymerized at 55°C for 12–24 h in gelatin capsules filled to exclude air (oxygen inhibits polymerization of acrylic resins). Spurr's resin is polymerized at 70°C. We routinely evacuate tissues in fresh resin at the polymerizing temperature to remove any small air bubbles, which improves the quality of the resulting sections. Reference to a standard text on electron microscopy (e.g., Hayat, 1970) should be made by the nonmicroscopist.
6. After polymerization of the resin, sections are cut by standard protocols using glass or diamond knives (for light and electron microscopy, respectively) fitted with a water boat.

For light microscopy, sections 2–3 μm thick need to be cut for there to be sufficient blue reaction product to be visible. However, such sections enable easy identification of individual cells. Sections are collected from the water surface with an eyelash or wire loop and dried onto a glass microscope slide. They should initially be examined without additional contrasting; subsequent staining with toluidine blue O (Feder and O'Brien, 1968) will assist interpretation of tissue structure, but will drastically reduce the contrast of the GUS reaction product. Aqueous toluidine blue (0.1% w/v) can be used for acrylic sections, while for epoxy sections the same stain (0.5% w/v) in benzoate buffer is used (Feder and O'Brien, 1968).

For electron microscopy, sections from 50 to 250 nm are cut on a diamond knife and collected on parlodion and carbon-coated grids by standard protocols (see Reed, 1974). Sections should initially be examined without contrasting and subsequently after staining with lead (e.g.,

Reynolds) and uranyl salts (2% aqueous for acrylic sections; saturated in ethanol for Spurr's sections). Acrylic sections are water miscible, and staining times of <1 min are required. For Spurr's sections, up to 10 min in stain will be required. GUS reaction product is crystalline and electron-dense, which makes identification easy.

Photography

Photography is possibly the most critical aspect of the histochemical assay for GUS activity. For whole-tissue photography, the specimen should be transferred to a medium of suitable refractive index such as immersion oil or glycerol, as described. Aqueous media can be used, but with inferior results. Careful attention should be paid to the optics and illumination. For a given specimen, the optimal combination of transmitted and/or incident light plus the background should be determined. Dark-field illumination can be extremely useful. For whole-mount photography, we use a Leitz MP 8 stereomicroscope equipped with an MPS 46 Photoautomat camera system, but small specimens can be viewed with considerable success on a compound light microscope. Sectioned material was examined with a Zeiss WL microscope equiped with Planapochromat optics.

Color images are recorded on Kodak Ektachrome professional 50 or 160 ASA film and black and white images are recorded on either Kodak Panatomic X or Polaroid Type 665 film. Electron micrographs are taken on a JEOL 100 S or 100 CX instrument operating at 60 or 80 kV using Kodak Fine Grain Release Positive 35 mm film or Kodak 4489 electron microscope sheet film.

Results and Discussion

The procedures described here have been successfully used, either in part or in full, to demonstrate GUS activity under the control of several promotors, including the CaMV 35S (Jefferson, 1987), the pea seed storage protein vicilin promotor (Higgins *et al.*, 1988), and hemoglobin promotors from *Parasponia* and *Trema* (Bogusz *et al.*, 1990). GUS activity has been observed in a range of transgenic tissues including mature seeds, seedlings, roots, leaves, and flowers. Of these, the only tissue with which difficulty was encountered was the tobacco flower, where browning due to polyphenol oxidase tended to mask fine detail of the GUS expression. Addition of polyvinyl pyrollidone (MW 40,000) may help prevent browning with such tissues (Flemming *et al.*, 1987).

Color Plate 5 and Figs. 1 and 2 show results obtained for whole tissue, cellular, and subcellular levels respectively. Color Plate 5 shows CaMV 35S-driven GUS expression in intact root tips of transgenic tobacco. The strongly stained area corresponds to the area of the root tip in which cells have little or no vacuole and are presumed to be very active metabolically. Many will be dividing. An area broadly corresponding to the vascular bundle is also stained, but the expression is weaker than in the tip cells. Neither the cells of the root cap nor the cortical cells proximal to the tip stain. No further detail can be obtained from such specimens, although optical sectioning with either differential interference contrast optics or a confocal microscope may be useful.

Figure 1a shows a root tip treated as per Color Plate 5, but then embedded in L R White and sectioned. Color images of such specimens reproduce poorly because of the lack of contrast. Each dark spot represents a blue reaction deposit, many of which encircle large, unstained areas that resemble vacuoles. In fact these unstained areas are nuclei

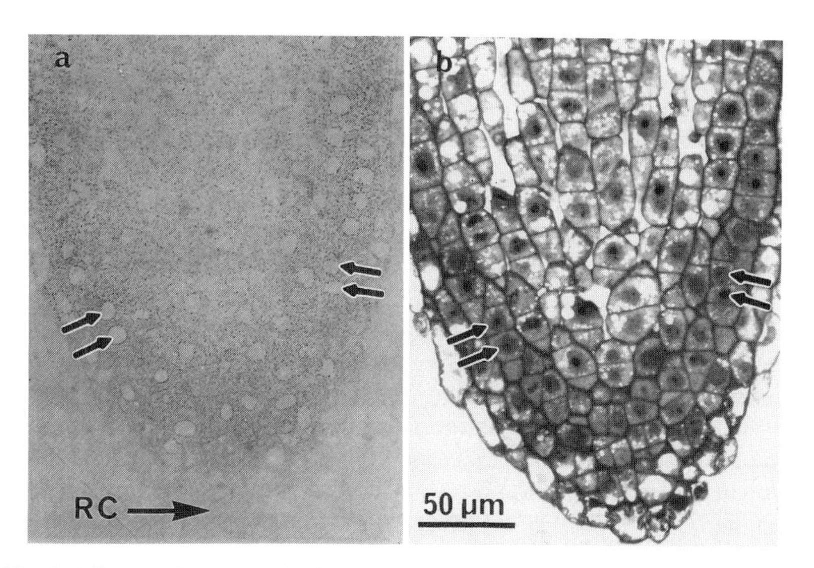

Fig. 1 Micrographs of root tip tissue from transgenic tobacco expressing a 35S GUS construct. Longitudinal section (2 μm thick) through the root tip: (a) without contrasting with toluidine blue, showing GUS reaction products as black deposits; (b) same section, following staining with toluidine blue. Arrows indicate four nuclei in the section. RC: root cap.

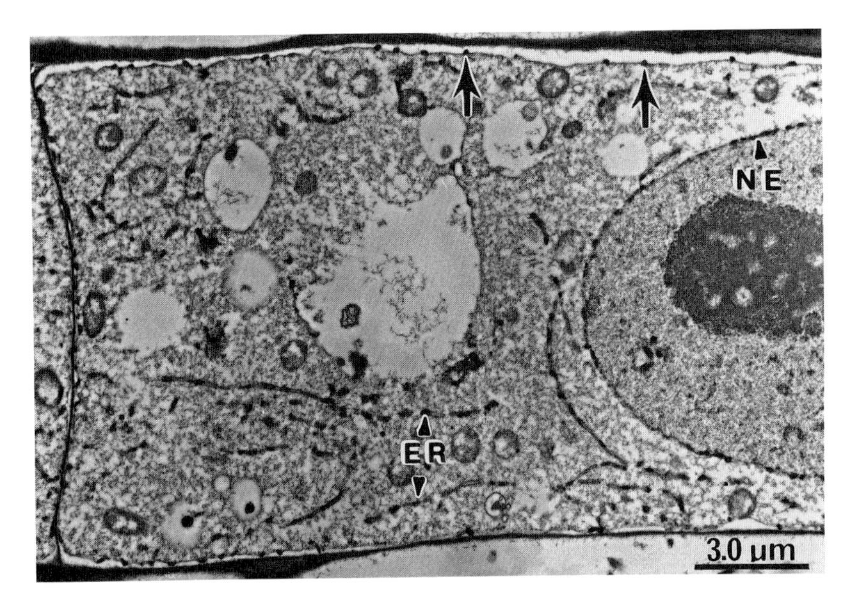

Fig. 2 Transmission electron micrograph of part of a provascular cell proximal to the meristematic region of the root tip. GUS reaction products are electron-dense and are present in the nuclear envelope (NE), the endoplasmic reticulum (ER), and at the plasma membrane (arrows).

that can be stained with toluidine blue O (Figure 1b). Following contrasting with toluidine blue, the GUS reaction products are difficult to identify.

From the same sample, ultrathin sections were cut and viewed in the transmission electron microscope (Figure 2), where the blue GUS reaction product deposits are visible without further treatment because of their electron density. In this tissue, GUS expression is driven by the 35S promotor and the reaction product is in the nuclear envelope (compare with Figure 1a) and the rough endoplasmic reticulum. It is also present at the plasmamembrane and in an unidentified cytoplasmic site(s). This pattern of distribution is consistent with the GUS enzyme being present in the endomembrane system, although positive identification of the Golgi membrane system has not been possible; presumably this membrane system is not preserved by the processing protocol.

The ability to detect GUS activity at the subcellular level, while an extremely powerful technique, needs to be viewed cautiously as the expression could also be influenced by fortuitous internal targeting sequences in the marker protein. This aspect will require further investigation.

Controls are an important part of GUS assays. Hu *et al.* (1990) described endogenous GUS-like activity, particularly in reproductive tissues of 11 out of 32 species examined. Therefore the histochemical GUS assay should always include untransformed tissue, or tissue transformed with a gene other than GUS. Kosugi *et al.* (1990) found that inclusion of methanol at up to 20% of the reaction volume reduced endogenous GUS-like activity to negligible levels. Besides false positive reactions, poor penetration of substrate can result in false negative results, although with the protocol used here, fixation renders membranes sufficiently permeable to overcome this potential problem. Dimethyl sulfoxide (DMSO) has also been used to facilitate penetration into tissues.

Formation of the blue reaction product is a two-step process, the initial product being colorless. The blue product results from oxidative dimerization of the indoxyl product; dimerization is enhanced by the potassium ferricyanide/ferrocyanide mixture, and for this reason we always include this in the substrate mixture.

Acknowledgments

The methods and observations described here were all made in collaborative studies with CSIRO colleagues Drs. E. J. Finnegan, E. S. Dennis, D. J. Llewellyn, W. T. Taylor, G. Ellis, and T. J. Higgins, who provided the transgenic material. The skilled technical assistance of Ms. Celia Miller is acknowledged.

References

Bogusz, D., Llewellyn, D. J., Craig, S., Dennis, E. S., Appleby, C. A., and Peacock, W. J. (1990). Nonlegume hemoglobin genes retain organ-specific expression in heterologous transgeninc plants. *Plant Cell* 2, 633–641.

Feder, N., and O'Brien, T. P. (1968). Plant microtechniques: some principles and new methods. *Am. J. Bot.* 55, 123–142.

Finnegan, E. J., Taylor, B. H., Craig, S., and Dennis, E. S. (1990).Transposable elements can be used to study cell lineages in transgenic plants. *Plant Cell* 1, 757–764.

Flemming, A. I., Wittenberg, J. B., Wittenberg, B. A., Dudman, W. F., and Appleby, C. A. (1987). The purification, characterisation and ligand-binding kinetics of hemoglobins from root nodules of the nonleguminous *Casuarina glauca-Frankia* symbiosis. *Biochim. Biophys. Acta* 911, 209–220.

Hayat, M. (1970). "Principles and Techniques of Electron Microscopy: Biological Applications." Van Nostrand, New York.

Hererra-Estrella, L., Depicker, A., Van Montague, M., and Schell, L. (1983). Expression of chimeric genes transferred into plant cells using a Ti-plasmid-derived vector. *Nature* 303, 209–213.

Higgins, T. J. V., Newbigin, E. J., Spencer, D., Llewellyn, D. J., and Craig, S. (1988). The sequence of a pea vicilin gene and its expression in transgenic tobacco plants. *Plant Mol. Biol.* 11, 683–695.

Hu, C-y., Chee, P. P., Chesney, R. H., Zhou, J. H., Miller, P. D., and O'Brien, W. T. (1990). Intrinsic GUS-like activities in seed plants. *Plant Cell Rep.* 9, 1–5.

Jefferson, R. A. (1987). Assaying chimeric genes in plants: The GUS gene fusion system. *Plant Mol. Biol. Rep.* 5, 387–405.

Jefferson, R. A., Burgess, S. M., and Hirsh, D. (1986). β-Glucuronidase from *Escherichia coli* as a gene fusion marker. *Proc. Natl. Acad. Sci. USA* 83, 8447–8451.

Kosugi, S., Ohashi, Y., Nakajima, K., and Arai, Y. (1990). An improved assay for β-glucuronidase in transformed cells: methanol almost completely suppresses a putative endogenous β-glucuronidase activity. *Plant Sci.* 70, 133–140.

Reed, N. (1974). Ultramicrotomy. *In* "Practical Methods in Electron Microscopy," Vol 3, part II (ed. A. M. Glauert). North Holland, Amsterdam.

PART 4

Applications of GUS to Plant Genetic Analysis

9 Review of the Use of the GUS Gene for Analysis of Secretory Systems

Leigh B. Farrell
Victoria, Australia

Roger N. Beachy
Department of Cell Biology
Scripps Research Foundation
La Jolla, California 92037

Introduction

While the *Escherichia coli* β-glucuronidase (GUS) gene has for some time been used as a reporter gene to investigate the spatial, temporal, and tissue-specific regulation of a variety of gene promoter sequences (Jefferson *et al.*, 1986; Jefferson, 1989) it has only recently been employed in organellar and membrane protein targeting studies. Kavanagh *et al.*(1988) demonstrated that the transit peptide and either 24, 53, or 126 amino acids of the mature chlorophyll *a*/*b* binding apoprotein direct GUS into a protease-resistant location within isolated *Nicotiana tobacum* chloroplasts. In the second documented use of GUS organellar targeting, GUS was delivered to the mitochondrial matrix in both yeast and tobacco mitochondria using presequences comprising the cleavable transit sequence and at least 67 amino acids of the mature yeast mitochondrial tryptophanyl-tRNA synthetase (Schmitz and Lonsdale, 1989). More recently, GUS has been imported into other organelles. By replacing six C-terminal amino acids of GUS with the C-terminal hexapeptide from the leaf peroxisomal enzyme glycolate oxidase, GUS was targeted to tobacco leaf peroxisomes (M. Volokita and P. Gonen, personal communication). Targeting of GUS into tobacco nuclei was afforded by fusion to the nuclear localization signal of mammalian SV40 large T-antigen or sequences encompassing the basic DNA-

GUS Protocols: Using the GUS Gene as a Reporter of Gene Expression

binding domain from the plant B-ZIP DNA binding protein TGA-1A (A. R. Van Der Krol and N-H. Chua, personal communication).

The utilization of GUS in targeting studies that include the endomembrane system has been complicated by the observation that while GUS could be targeted to the endoplasmic reticulum of tobacco, the enzymatic activity of GUS was severely inhibited due to *N*-glycosylation (Iturriaga *et al.*, 1989). Inspection of the predicted amino acid sequence for GUS, derived from the published nucleotide sequence (Jefferson *et al.*, 1986) of the *gusA* gene, revealed potential *N*-glycosylation acceptor sites at amino acid positions 358 (AsnLeuSer) and 423 (AsnLeuSer) (Iturriaga *et al.*, 1989). To extend the use of GUS as a reporter to endomembrane targeting studies, a program was initiated to sequentially destroy each of the putative *N*-glycosylation sites using oligonucleotide mediated site-directed mutagenesis (Farrell and Beachy, 1990). Oligonucleotides were designed for site-directed mutagenesis and used to destroy both of these potential *N*-glycosylation acceptor sites by substituting asparagine codons in these sites with either serine, threonine, or proline. Putative mutants were first screened directly for GUS enzymatic activity in *E. coli* by the expression of an in-frame fusion of *gusA* with the *lacZ* gene on the plasmid pBluescript KS+ (Stratagene, Inc., La Jolla, Calif.) and secondly at the nucleotide level to confirm that the correct mutations had been introduced into the GUS coding region. GUS substitution mutants were recovered for substitutions in the putative *N*-glycosylation acceptor site at codon position 358 but not for codon position 423. Subsequent nucleotide sequence analysis of sequences around codon 423 identified one proximal and two distal nucleotide insertions relative to this site that cause a translational frameshift that leads to the substitution of amino acids from codons 420 to 425 and the insertion of one amino acid at codon 426: thus the putative *N*-linked glycosylation site identified in the deduced amino acid sequence of the published GUS gene sequence does not exist.

Of the three amino acid substitutions made at position 358, the serine substitution was the most conservative, with 64% of wild-type GUS activity in extracts from transformed *E. coli* cells, followed by threonine (16%) and proline (0.5%). The plasmid carrying this clone in plasmid pUC119 has been designated pGUS : Asn-Ser and is available from Clontech, Inc., Palo Alto, Calif.

To demonstrate that the GUS : Asn-Ser substitution mutant is a suitable reporter for endomembrane targeting studies, GUS : Asn-Ser was fused to the endomembrane targeting sequences from the α subunit of the soybean seed storage protein β-conglycinin and introduced into plants. Soybean β-conglycinin is localized to seed protein bodies and comprises various combinations of three subunits: α' (76 kDa), α (72 kDa), and β (53 kDa), each of which is synthesized as a larger precursor. Maturation of the α subunit precursor to the form found in seed protein bodies involves two proteolytic cleavage events: one involves the cotranslational cleavage of approximately 22 amino acids during translocation into the endoplasmic reticulum (Sengupta *et al.*, 1981), and the other involves the posttranslational cleavage of a further 40 amino acids, presumably in protein bodies. In addition, the α subunit is *N*-glycosylated at two sites during passage through the endomembrane system.

We report results of experiments that demonstrate that GUS : Asn-Ser gives a 5- to 10-fold enhancement of enzymatic activity relative to wild-type GUS when targeted to the endoplasmic reticulum. This should enable GUS to be used as a suitable reporter for the investigation of protein targeting studies involving the secretory pathway.

Evaluation of GUS: N358-S as a Reporter for Secretory Protein Targeting Studies

To determine whether the glycosylation resistant version of GUS gives higher activity than that of wild-type when targeted to the endoplasmic reticulum, the first 34 and 62 N-terminal amino acids of the α subunit signal sequence coding region were fused to both wild-type and glycosylation-resistant versions of the GUS gene. These chimeric genes were transformed into *Nicotiana benthamiana,* via *Agrobacterium tumefaciens* on the binary plant expression vector pα'C8/505, which comprises the seed specific promoter and 3'-untranslated sequences from the α' subunit of β-conglycinin in pMON505 (Horsch and Klee, 1986) (Figure 1).

When the GUS activity in extracts from transgenic seed was measured (Jefferson, 1987), the glycosylation-resistant mutant of each pair had at least 5- to 10-fold greater activity than the corresponding wild-type construct (Figure 2A). It is important to note that when seed extracts were centrifuged prior to withdrawing aliquots for GUS assays, a two- to threefold reduction in GUS activity was observed for all constructs including nonfused GUS (data not shown). These data would indicate that GUS protein associates with particulates in the cell debris. For the present assays, extracts were not centrifuged prior to completing GUS assays.

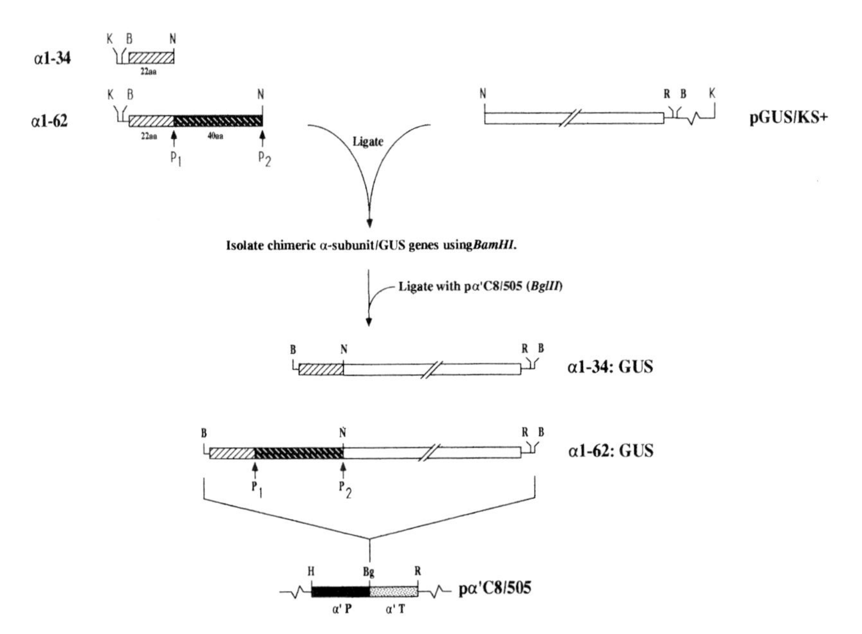

Fig. 1 Construction of α subunit : GUS fusions. The α subunit gene contained in pUC119 (Vieira and Messing, 1987) was modified using site-directed mutagenesis to introduce *NcoI* (N) restriction enzyme recognition sites at either codon position 34 or 62. Sequences encoding the first 34 (22 amino acid signal sequences plus 12 amino acids of the pro-sequence) or 62 (22 amino acid signal sequence plus entire 40 amino acid pro-sequence) N-terminal amino acids of the α subunit were excised from these plasmids using *KpnI* (K) and *NcoI*. These fragments were ligated at the *KpnI* and *NcoI* sites of plasmids that carried both wild-type and mutant GUS genes in the plasmid pBluescript KS+ (Stratagene, Inc.). Fusion of the α subunit sequences with GUS introduces one additional amino acid at the fusion site. Chimeric α subunit : GUS gene fusions were mobilized using *BamHI* (B) and ligated into the *BglII* (Bg) site of the binary plant seed-specific expression vector pα'C8/505, which comprises the promoter and 3′-untranslated sequences from the α' subunit gene carried in the binary vector pMON505. Other abbreviations: P_1, first proteolytic cleavage site; P_2, the second proteolytic cleavage of the α subunit precursor; R, *EcoRI;* and H, *HindIII*.

When mobilities of fusion proteins were compared by sodium dodecyl sulfate (SDS) polyacrylamide gel electrophoresis and Western immunoblots, each of the wild-type constructs had a retarded mobility compared to that of the corresponding glycosylation resistant construct (Figure 2B). The size difference of approximately 2 kDa between corresponding wild-type and mutant fusion proteins is consistent with the addition of an N-linked oligosaccharide during the passage of wild-type GUS through the endomembrane system. No such shift in mobility was

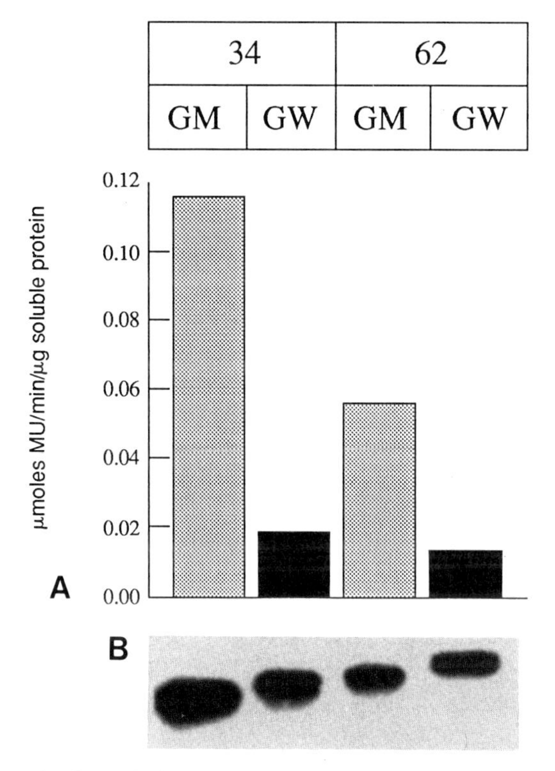

Fig. 2 Characterization of fusion proteins between the α subunit and either wild-type or mutant GUS. (A) GUS activity of fusion proteins in *N. benthaminana* seed extracts. Approximately 5 mg of dry seed was homogenized in 250 μl of GUS extraction buffer (Jefferson, 1987). Following centrifugation of the homogenate, aliquots of the soluble fraction were assayed for GUS activity using the standard fluorometric assay (Jefferson, 1987). The concentration of soluble protein in seed homogenates was estimated by the method of Bradford (1976). GUS enzymatic activity is expressed as μmol MU/min/μg soluble protein. Histogram header abbreviations: 34 and 62 refer to the number of amino acids of the α subunit that are fused to either mutant (GM) or wild-type (GW) GUS. (B) Western analysis of α subunit : GUS fusion proteins. Approximately 50 μg of soluble protein from seed extracts was fractionated on an 8% SDS-polyacrylamide gel and Western blotted to nitrocellulose. Blots were then reacted with rabbit antisera directed against GUS (Clontech, Inc., Palo Alto, Calif.) and then ^{125}I-labeled donkey anti-rabbit $(Fab)'_2$ antibodies. GUS fusion proteins were visualized by autoradiography of the blots.

observed when the mobilities of wild-type and mutant GUS lacking a signal peptide were compared (Farrell and Beachy, 1990). Similar results with respect to both GUS activity and gel mobility were obtained when fusion proteins were expressed under the control of the cauliflower mosaic virus (CaMV) 35S promoter (data not shown).

Perspectives

The substitution of the asparagine residue at position 358 for serine prevents glycosylation of the protein in the endoplasmic reticulum thereby retaining enzymatic activity. Thus, the advantages of the GUS system that have been exploited in studies of protein targeting to mitochondria, chloroplasts, peroxisomes, and nuclei can now be extended to protein targeting studies involving the secretory pathway.

As a continuation of the present study, GUS : Asn-Ser is now being utilized as a reporter to map vacuolar targeting domains engendered within the α subunit of β-conglycinin. While it is established that an N-terminal hydrophobic signal sequence is required for translocation of the α subunit across the endoplasmic reticulum (Sengupta *et al.*, 1981), additional topological signals are assumed to be requisite for vacuolar targeting (Rothman, 1987; Wieland *et al.*, 1987; Dorel *et al.*, 1989; Denecke *et al.*, 1990; Bednarek *et al.*, 1990). To delineate the vacuolar targeting sequences within the α subunit, chimeric genes encoding a series of C-terminally truncated sequences from the α subunit fused to GUS : Asn-Ser have been introduced into plants using the *Agrobacterium tumefaciens* method of transformation. Analysis of fusion proteins expressed in transgenic plants indicates that GUS : Asn-Ser can be fused to large N-terminal extensions of at least 600 amino acids while retaining significant enzymatic activity (L. B. Farrell, unpublished data). At present those sequences in the α subunit that target GUS to vacuoles are being determined by a combination of subcellular fractionation and electron microscopy of both immunogold-labeled seed tissue sections and fixed tissue sections that have been reacted with 5-bromo-3-indolyl glucuronide (X-Gluc, Jersey Lab Supply, Livingston, New Jersey; for protocol, see Craig, Chapter 8).

Two approaches have been evaluated for subcellular fractionation. The first involves the fractionation of protein bodies from mature tobacco seeds using potassium iodide/glycerol gradients (Sturm *et al.*, 1988). In this fractionation system, seeds are homogenized in anhydrous glycerol (1.26 g/ml) and then fractionated by ultracentrifugation through a step gradient comprising two steps of 1.3 g/ml and 1.4 g/ml potassium iodide/glycerol steps. Following centrifugation, dense protein bodies are collected at the interphase between the 1.3 g/ml and 1.4 g/ml steps and dialyzed against GUS extraction buffer (Jefferson, 1987) to remove potassium iodide and glycerol, which drastically inhibit GUS enzymatic activity. This method would appear not to be useful for the identification of sequences that target GUS to protein bodies. When seed from transgenic tobacco plants, which express non-fused GUS under the control of the α' subunit gene promoter, was fractionated using this system, GUS activity was found mainly in the heavy protein body fraction. The specificity of this interaction is uncertain; however, it should be noted that electron microscopic analysis indicated that the heavy protein body fraction contained significant amounts of cell wall debris (M. Bustos, personal communication). Given the observation that GUS enzyme activity is also associated with the pellet in seed extracts in the presence of GUS extraction buffer, it is possible that GUS associates with plasma membrane or cell wall components in the potassium iodide/glycerol fractionation system. As an alternative to this method, a second fractionation method for purifying tobacco leaf vacuoles, which are homologs of protein bodies (more recently termed seed protein storage vacuoles) is being evaluated. This method is a modification of a method used to purify vacuoles from *Pisum sativum* (Guy *et al.*, 1979; Wilkins *et al.*, 1990; Bednarek *et al.*, 1990). In principle, protoplasts, released from leaves of axenically grown tobacco plants by digestion with cellulase and macerozyme, are fractionated through a two-step gradient comprising 5 and 10% Ficoll. During centrifugation, vacuoles, released from cells by shearing forces, float to the surface of the gradient. The vacuole fraction is recovered and adjusted to 10% Ficoll and recentrifuged as before. After this second centrifugation, vacuoles are essentially void of cytoplasmic contamination as judged by measurement of the activities of vacuolar marker enzymes (aryl-α-mannosidase and acid phosphatase) and the pseudo-cytoplasmic marker catalase, which is released from fragile peroxisomes during centrifugation (Wilkins *et al.*, 1990).

Acknowledgments

We thank Jacqueline DePaulo for her excellent technical assistance, Donna Droste for plant transformation, Mike Dyer for management of transgenic plants, and Philip Lessard for critically reviewing this manuscript. The work reported here has been supported by the Monsanto Company and a grant from the Missouri Research Assistance Act.

References

Bednarek, S. Y., Wilkins, T. A., Dombrowski, J. E., and Raikhel, N. V. (1990). A carboxyl-terminal propeptide is necessary for proper sorting of barley lectin to vacuoles of tobacco. *Plant Cell.* 2, 1145–1155.

Bradford, M. (1976). A rapid and sensitive method for the quantities of protein utilizing the principle of protein-dye binding. *Anal. Biochem.* 72, 248–254.

Denecke, J., Botterman, J., and Deblaere, R. (1990). Protein secretion in plant cells can occur via a default pathway. *Plant Cell* 2, 51–59.

Dorel, C., Voelker, T. A., Herman, E. M., and Chrispeels, M. J. (1989). Transport of proteins to the plant vacuole is not by bulk flow through the secretory system, and requires positive sorting information. *J. Cell Biol.* 108, 327–337.

Farrell, L. B., and Beachy, R. N. (1990). Manipulation of β-glucuronidase for use as a reporter in vacuolar targeting studies. *Plant Mol. Biol.* 15, 821–825.

Guy, M., Reinhold, L., and Michaili, D. (1979). Direct evidence for a sugar transport mechanism in isolated vacuoles. *Plant Physiol.* 64, 61–64.

Horsch, R., and Klee, H. (1986). Rapid assay of foreign gene expression in leaf discs transformed by *Agrobacterium tumefaciens:* Role of T-DNA borders in the transfer process. *Proc. Natl. Acad. Sci. USA* 83, 4428–4432.

Iturriaga, G., Jefferson, R. A., and Bevan, M. W. (1989). Endoplasmic reticulum targeting and glycosylation of hybrid proteins in transgenic tobacco. *Plant Cell* 1, 381–390.

Jefferson, R. A. (1987). Assaying chimeric genes in plants. The GUS fusion system. *Plant Mol. Biol. Rep.* 1, 387–405.

Jefferson, R. A. (1989). The GUS reporter gene system. *Nature* 342, 837–838.

Jefferson, R. A., Burgess, S. M., and Hirsh, D. (1986). β-Glucuronidase from *Escherichia coli* as a gene fusion marker. *Proc. Natl. Acad. Sci. USA* 83, 8447–8451.

Kavanagh, T. A., Jefferson, R. A., and Bevan, M. W. (1988). Targeting a foreign protein to chloroplasts using fusions to the transit peptide of a chlorophyll a/b protein. *Mol. Gen. Genet.* 215, 38–45.

Rothman, J. E. (1987). Protein sorting by selective retention in the endoplasmic reticulum. *Cell* 50, 521–522.

Schmitz, U. K., and Lonsdale, D. M. (1989). A yeast mitochondrial presequence functions as a signal for targeting to plant mitochondria in vivo. *Plant Cell.* 1, 783–791.

Sengupta, C. Deluca, V., Bailey, D., and Verma, D. P. S. (1981). Post-translational processing of 7S and 11S components of soybean storage proteins. *Plant Mol. Biol.* 1, 19–34.

Sturm, A., Voelker, T. A., Herman, E. M., and Chrispeels, M. J. (1988). Correct glycosylation, Golgi-processing, and targeting to protein bodies of the vacuolar protein phytohemagglutinin in transgenic tobacco. *Planta* 175, 170–183.

Vieira, J., and Messing, J. (1987). Production of single-stranded plasmid DNA. *Methods Enzymol.* 153, 3–11.

Wieland, F. T., Gleason, M. L., Serafina, T. A., and Rothman, J. E. (1987). The rate of bulk flow from the endoplasmic reticulum to the cell surface. *Cell* 50, 289–300.

Wilkins, T. A., Bednarek, S. Y., and Raikhel, N. V. (1990). Role of propeptide glycan in post-translational processing and transport of barley lectin to vacuoles in transgenic tobacco. *Plant Cell* 2, 301–313.

10 Applications of GUS to Molecular Plant Virology

Jane K. Osbourn
Department of Virus Research
John Innes Institute
AFRC Plant Science Research Center
Norwich, United Kingdom

T. Michael A. Wilson
Center for Agricultural
Molecular Biology
Rutgers University
New Brunswick, New Jersey

The exquisite sensitivity and low background expression levels of the β-glucuronidase (GUS) gene system are well suited to molecular studies on virus–plant cell interactions. For the past 6 years, plant RNA viruses, which comprise the huge majority (~94%) of all known plant viruses (~650), have been amenable to studies by recombinant DNA techniques through the construction of full-length cDNA clones and the use of efficient *in vitro* transcription plasmids (Ahlquist, 1986). The use of virus-based, autonomous cytoplasmic replicons can provide very high levels of selected gene expression. For reasons of chronology, the first reporter gene inserted into chimeric plant viral RNAs was not GUS, but chloramphenicol acetyltransferase (CAT) (French *et al.*, 1986; Takamatsu *et al.*, 1987). GUS would now seem to be the "foreign" gene of choice; however, more recently published work on single-stranded DNA geminiviruses as gene amplification and expression vectors in plants again used either CAT or neomycin phosphotransferase (*neo*) as reporter genes (Ward *et al.*, 1988; Hayes *et al.*, 1988). In part, the latter may reflect gene size-limited packaging constraints for efficient virus spread. The ability to follow directly the spread of the recombinant viral genome from cell to cell, by virtue of GUS gene expression *in situ* using histochemical substrates, is extremely attractive experimentally.

To detect the primary events of gene expression *in vivo* using biologically realistic input doses of virus inoculum (the "parental" virus parti-

GUS Protocols: Using the GUS Gene as a Reporter of Gene Expression

cles), or of viruslike RNAs, requires either the high sensitivity of the GUS gene assay, or polymerase chain reaction (PCR) and/or RNA toeprinting techniques, or highly radioactive virions and/or nascent polypeptide products. All these approaches have been, or are being, used by those engaged in studies on the so-called "early events" of plant virus infection. Below we describe a novel technology for creating and utilizing GUS mRNP structures for such purposes, and for elucidating the mechanism(s) behind genetically engineered protection in transgenic plants expressing a viral coat protein (CP) gene. Some years ago we used similar TMV-like pseudovirus particles, but containing CAT mRNA, to visualize the site of nucleocapsid disassembly *in vivo* (Plaskitt *et al.*, 1988). Today, GUS particles and histochemical staining could be used to even better effect.

Another application of GUS gene expression that we have described, within the broader context of plant virus–plant cell phenomena, concerns the translational enhancement of gene expression in general, and in isolated plant cells, protoplasts, or transformed plants in particular, using a *cis*-active 5′-leader sequence from a plant RNA virus.

Enhancement of GUS mRNA Expression in Plant Systems

Extensive series of experiments have been performed to document the effect of the 5′-untranslated leader sequence from genomic TMV RNA, or other viral RNA leaders, on translation of a wide variety of prokaryotic and eukaryotic reporter mRNAs in a range of cell-free and intact cell expression systems. In some we elected to use GUS mRNAs with either the native AUG context or a mutagenized AUG context (pRAJ275; Sleat *et al.*, 1987; Jefferson, 1987).

GUS mRNAs with or without the particular 5′-leader sequence of interest were transcribed *in vitro* from SP6 or T7 RNA-polymerase promoter plasmids, and were added directly to either plant or animal cell-free translation systems (here expression was monitored only by incorporation of L-[^{35}S]methionine into GUS protein). Alternatively, transcripts were electroporated into tobacco mesophyll protoplasts (Table 1) (Gallie *et al.*, 1987a) or a variety of other plant protoplasts (Gallie *et al.*, 1989). In some experiments, chimeric leader–GUS

Table 1

Translational Enhancement by Various Viral RNA Leaders Attached to Native AUG-Context GUS mRNA and Electroporated into Tobacco Mesophyll Protoplasts[a]

Leader	Relative[b] GUS activity in extracts
Native GUS leader	1.0
TMV RNA (U1) = Ω′	18.0
TMV RNA (SPS) = Ω′	14.0
AlMV RNA4 leader	8.0
BMV RNA3 leader	8.0
Rous sarcoma genomic leader	8.0
TYMV genomic leader	<0.3

[a] See Gallie *et al.* (1987a). AlMV = alfalfa mosaic virus; BMV = brome mosaic virus; TYMV = turnip yellow mosaic virus; TMV = tobacco mosaic virus (U1 or common strain; SPS or tomato strain).

[b] Relative specific activities (nmol MUG hydrolyzed/min/μg leaf protein) expressed as viral leader GUS mRNA/native GUS mRNA.

mRNAs were expressed *in situ* in transformed bacterial cells (Gallie *et al.*, 1987a; Gallie and Kado, 1989) from an inducible *trp* promoter on a pSP64(Promega)-derived plasmid (pSP64TMV; pJII1; Sleat, 1987; Gallie *et al.*, 1987b), or in plant protoplasts from linearized pTZ18-derived plasmid DNAs bearing a cauliflower mosaic virus (CaMV) 35S promoter (Wilson *et al.*, 1990). In all the latter cases, GUS gene expression was monitored either spectrophotometrically (bacterial extracts; *p*-nitrophenyl-β-D-glucuronide) or spectrofluorimetrically [plant extracts; 4-methyl-umbelliferyl-β-D-glucuronide (MUG)]. Systematic experiments are in progress (M. J. Dowson-Day, personal communication) or have been completed but not yet published (V. Buchanan-Wollaston, personal communication) using GUS gene transcripts with or without the TMV or other leader sequence to measure translational enhancement of GUS expression in stably transformed tobacco plants. The results so far largely confirm earlier findings from transient expression assays, showing a 5- to 10-fold enhancement of GUS activity when the TMV leader sequence is present.

Use of GUS to Probe the Mechanism(s) of Coat Protein-Mediated Protection against Plant Viruses

Since 1986, a variety of stable, *Agrobacterium*-transformed plants expressing a particular viral coat protein (CP) gene have been shown to resist challenge inoculations with the virus corresponding to the specific CP which they express, or to closely related viruses with >60% CP homology (Stark and Beachy, 1989; Nejidat and Beachy, 1990). Recently, simultaneous resistance to potato viruses X and Y was achieved by creating double CP-transgenic potato plants (Lawson *et al.,* 1990). Crop plants not amenable to transformation by *A. tumefaciens* (e.g., monocotyledonous species) can be transformed by suitable DNA delivered via a "particle gun" (Sanford, 1988); however, in these cases subsequent protocols for tissue culture and plant regeneration are not trivial.

Examples of other CP-virus systems in addition to TMV (Powell Abel *et al.,* 1986; Nelson *et al.,* 1987; Register and Beachy, 1988) include cucumber mosaic virus (CMV; Cuozzo *et al.,* 1988), potato virus X (PVX; Hemenway *et al.,* 1988; Hoekema *et al.,* 1989), alfalfa mosaic virus (AlMV; Loesch-Fries *et al.,* 1987; Tumer *et al.,* 1987), tobacco rattle virus (TRV; van Dun and Bol, 1988), and tobacco streak virus (TSV; van Dun *et al.,* 1988). Clearly, many other systems are under scrutiny at the moment.

The mechanism(s) of CP-mediated protection remain unclear, but a number of characteristics of the process have now been identified. Protection is dependent on the relative concentrations of endogenous CP expressed by the plant and of the virus used as challenge inoculum (Nejidat and Beachy, 1989). Tobacco plants transformed to express TMV CP resist concentrations of mechanically inoculated virus up to about 1–10 μg/ml. In almost all the cases above, the CP^+ transgenic plants did not show substantial resistance to naked or partially uncoated homologous viral RNA, and no real protection against unrelated viruses or viral RNAs. One exception was PVX CP^+ plants challenged with PVX RNA (Hemenway *et al.,* 1988). Protection is also effective at the protoplast level (Register and Beachy, 1988, 1989) and so cannot be due only to inhibition of viral cell-to-cell spread functions. To investigate the mechanism of CP-mediated protection requires identifying stages of the virus replication cycle that could be inhibited by free CP.

TMV is mechanically transmitted and enters plant cells through damaged cell walls and membranes (Plaskitt *et al.*, 1988). Once inside the cell, virions probably disassemble cotranslationally (Shaw *et al.*, 1986) by analogy with events discovered *in vitro* (Wilson, 1984). A cytoplasmic 40S ribosomal subunit binds to the exposed 5′-end of the viral RNA (Wu *et al.*, 1984; Mundry *et al.*, 1991), scans to the start codon (Kozak, 1989), binds a 60S subunit, and initiates translation. During the first 80S translocation event along the viral RNA, CP is removed progressively. In the absence of virus-specific cell receptors, parental (inoculum) virus disassembly is considered the first stage of infection, which could be inhibited in CP^+ transformed plants, especially since CP-mediated protection is sensitive both to the level of CP-gene expression and to the concentration of challenge virus, and since it is overcome by partially or completely naked viral RNA.

To investigate this hypothesis we created nonreplicating "pseudovirus" particles (Sleat *et al.*, 1986). Pseudovirus particles contain a convenient reporter mRNA (e.g., GUS), usually with an efficient 5′-leader sequence (e.g., the TMV genomic leader Ω, see above) and a 3′-proximal "origin-of-assembly sequence" (OAS); the chimeric transcript is then packaged in viral (TMV) CP *in vitro*. Structurally, pseudovirus particles look and behave like virions. They are able to uncoat and express their nonreplicating "pseudogenomes" transiently and often very efficiently *in vivo*, presumably by cotranslational disassembly (Plaskitt *et al.*, 1988; Gallie *et al.*, 1987c), and they encode an assayable reporter enzyme (CAT or GUS) in place of a viral nonstructural protein. For a given construct therefore, the extent of reporter gene expression reflects the number of pseudovirus particles disassembled. Thus the relative levels of reporter (GUS) gene expression in pseudovirus-inoculated protoplasts prepared from control (CP^-) tobacco or from CP^+ transgenic plants gives a measure of the extent of inhibition of viruslike particle uncoating by the endogenous CP (Osbourn *et al.*, 1989).

A bacteriophage SP6 promoter transcription plasmid (pJII140; Figure 1) (Sleat *et al.*, 1987) containing (sequentially) Ω′, a GUS gene (from pRAJ275; Jefferson, 1987), a 440-nucleotide sequence encompassing three hairpin loops of the TMV OAS (Sleat *et al.*, 1986), and finally a convenient, unique restriction site for plasmid linearization was used to generate GUS mRNA with suitably functional termini. Encapsidation by TMV CP *in vitro* (Jupin *et al.*, 1989) resulted in extremely stable TMV-like rods 18 nm in diameter but only 112 nm in length (Figure 2; full-size TMV = 300 nm). After packaging, pseudovirus

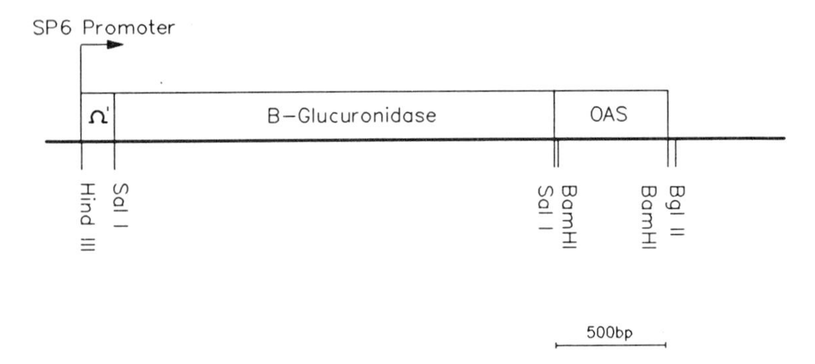

Fig. 1 Map of SP6 transcription plasmid pJII140 (Sleat *et al.*, 1987). pJII140 was derived from pSP64 (Promega Corp.) via pSP64TMV (Sleat *et al.*, 1986), which contained a *Bam*HI fragment bearing the TMV origin-of-assembly sequence (OAS; genome coordinates 5118-5550). The GUS gene contains no common restriction sites (Jefferson, 1987). *Bgl*II-linearized pJII140 was transcribed as described in the text prior to incubation with TMV coat protein.

preparations are treated extensively with micrococcal nuclease to degrade any residual unencapsidated or partially encapsidated RNA.

The cooperative, bidirectional self-assembly mechanism of TMV has permitted an almost universal length- and sequence-independent RNA packaging system to be developed (Sleat *et al.*, 1986; Jupin *et al.*, 1989). So far, only ribosomal RNAs and ssDNA have failed to become encapsidated (Sleat *et al.*, 1986; Gallie *et al.*, 1987d).

Pseudovirus particles can be used to probe the early structural events of virus infection. They can be introduced into tobacco protoplasts by polyethylene glycol treatment, or by electroporation, and the transient level of reporter gene expression can be used to assess the extent of nucleocapsid disassembly without the complication of progeny RNA formation and expression.

If CP-mediated protection depends on inhibited virus uncoating, then one would expect to see high (GUS) gene expression in pseudovirus-inoculated control (CP^-) tobacco protoplasts compared to the level of expression in similarly treated CP^+ transgenic protoplasts. Results from such an experiment are shown in Table 2. Mock-inoculated protoplast samples of either type are used to provide a measure of any (low) endogenous GUS activity. Unencapsidated GUS mRNA was also added to protoplasts to assess the relative efficiency and variability of

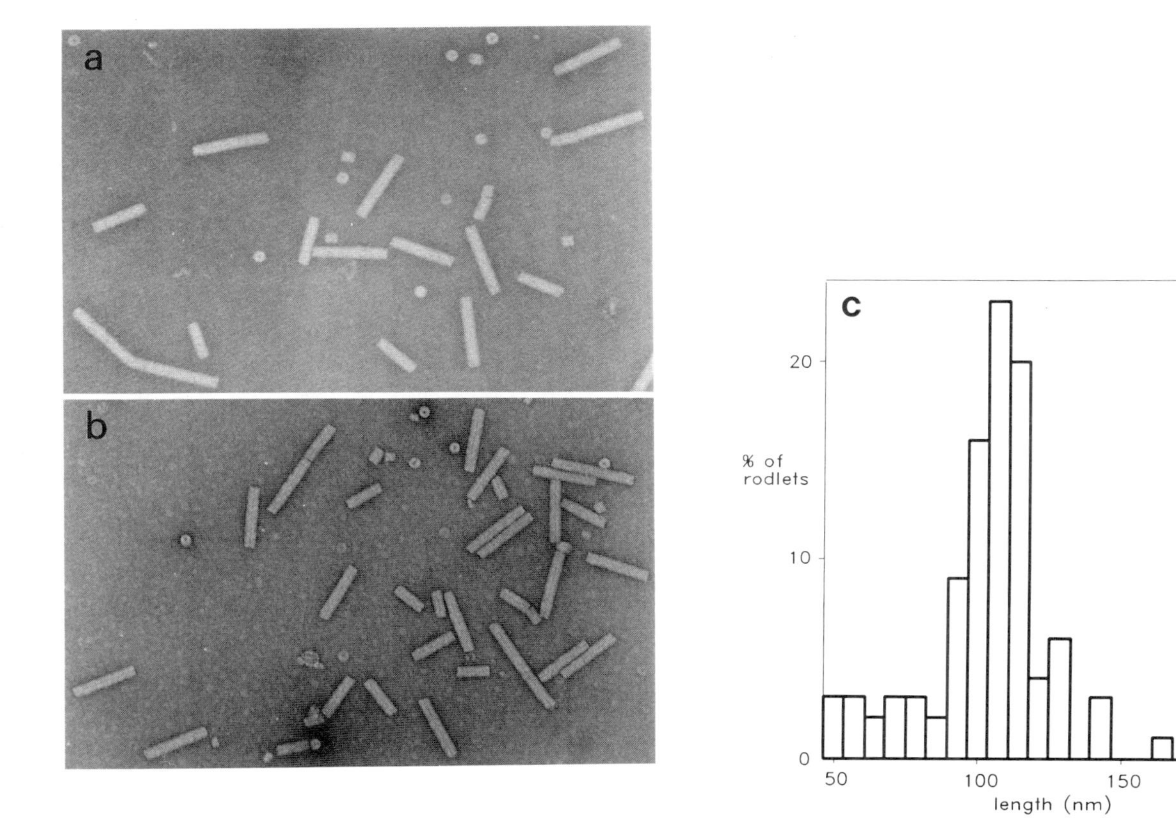

Fig. 2 Electron micrographs of pseudovirus particles containing Ω′-GUS-mRNA (2.4 kb) transcribed from *Bgl*II-cut pJII140 (Fig. 1). Two independent preparations of GUS rods are shown: (a) directly packaged (Jupin *et al.*, 1989) or (b) packaged after nucleic acid purification as described. Transcript length predicts rodlets of 112 nm, the mode of the frequency-length histogram (c) (6.4 kb TMV RNA = 300 nm).

Table 2

GUS Activities in Control or CP$^+$ Transgenic Tobacco Mesophyll Protoplasts Electroporated with Naked or Encapsidated GUS mRNA ("Pseudovirus")[a]

	GUS activity (pmol 4-MU[b]/min/mg total soluble protein)		
Inoculum	Control *Xanthi* protoplasts	(U1) CP$^+$ transgenic protoplasts	Ratio[c]
Mock	<6	<6	—
Naked GUS mRNA	169	61	2.8
Packaged GUS mRNP	14,900	152	98.0

[a] See Osbourrn *et al.* (1989).
[b] 4-Methylumbelliferone (fluorescent product).
[c] Calculated as GUS activity in control protoplast extract/GUS activity in CP$^+$ transgenic protoplast extract.

the electroporation technique, as well as the quality and viability of the two independent classes of protoplasts. Some of the 2.8-fold reduction in naked GUS mRNA expression in CP$^+$ protoplasts (Table 2) may reflect biological or experimental variation. However, since Register and Beachy (1988) reported that some CP-mediated protection was detectable in analogous experiments with 4 μg/ml TMV RNA or less, some of the 2.8-fold reduction may also arise through endogenous CP interacting with either the TMV-derived Ω' or OAS sequences in the chimeric GUS mRNA. Nevertheless, with encapsidated GUS mRNA there was a 98-fold reduction in GUS activity in CP$^+$ transgenic protoplasts compared to control tobacco protoplasts (Table 2). Thus it appears that CP$^+$ transgenic protoplasts are relatively inefficient at uncoating GUS mRNA contained in the homologous CP, implying that one significant mechanism of CP-mediated protection involves substantial inhibition of parental virus particle disassembly, thereby reducing the effective genome inoculum concentration.

As well as compatible CP–CP interactions, the importance of 5′-leader sequence interactions with the endogenous CP can also be assayed using the GUS–pseudovirus system. Different artificial or viral leader–GUS–OAS constructs have been made and assembled into pseudovirus particles. By electroporating these into the two classes of tobacco protoplasts described above, direct leader effects on nucleoprotein particle stability and GUS mRNA translation can be resolved from CP-mediated interference phenomena at the leader. Parallel experiments *in vitro*, using GUS pseudovirus particles with different leader sequences translated in the presence or absence of excess free TMV CP, indicate that the relative inhibition of cotranslational disassembly (measured by incorporation of L-[^{35}S]methionine into GUS protein) is greatest when the native TMV leader sequence (Ω) is present than with other unrelated leaders. Analogous experiments were performed several years ago (Wilson and Watkins, 1986) to show that excess TMV CP selectively inhibited the cotranslational disassembly of TMV particles compared to naked RNA; however, some inhibition of the latter was always evident. These data (J. K. Osbourn, unpublished results) suggest that native TMV leader–CP interactions may also be important to inhibit cotranslational disassembly during CP-mediated protection in transgenic tobacco cells. Future research will be directed toward electroporating GUS pseudovirions with different 5′-leader sequences into CP$^+$ transgenic or control tobacco protoplasts to assay the relative transient levels of GUS mRNA expression.

In tobacco protoplasts, GUS is a more sensitive reporter gene system than CAT. We, and others, have found that tobacco plants contain nonspecific enzymes that mimic CAT activity by acetylating some (usually $<1\%$) of the substrate, [^{14}C]chloramphenicol. This resulted in an unacceptably high background "CAT activity," measurable even before CAT pseudovirus particles were introduced into CP$^+$ or control protoplasts. Immunocytochemical labeling of control tobacco leaf sections with polyclonal rabbit antiserum raised against CAT protein (a generous gift from Dr. W. V. Shaw, Leicester University, UK) failed to show any gold labeling (Plaskitt *et al.*, 1988), confirming that the contaminating tobacco enzyme activity is not immunologically related to CAT.

In conclusion, GUS pseudovirus particles in particular provide a rapid and convenient tool to assay for viruslike nucleocapsid disassembly *in vivo*, in control and genetically engineered "protected" plants, cells, or protoplasts. It will be interesting to introduce such structures into transformed protoplasts that express other viral CP genes. Alternatively, plants expressing other forms of genetically engineered virus resistance such as antisense RNA, 3′- viral-sense RNA, self-cleaving ribozymes, or expression of non-structural viral proteins (e.g., the TMV 54-kDa polypeptide: Dr. M. Zaitlin, Cornell University, N.Y., personal communication) can be screened for their effect (or otherwise) on ribonucleocapsid stability.

Protocols

mRNA Production

1. Synthesize transcripts from *Bgl*II-cut plasmid pJII140 (Fig. 1) (Sleat *et al.*, 1987) using SP6 RNA polymerase as described (Melton *et al.*, 1984), except omit the bovine serum albumin and use a 50-μl reaction volume.
2. After 2 h, phenol/chloroform extract the reactions twice and chloroform extract once. Recover total nucleic acid by ethanol precipitation at −20°C for 18 h and centrifugation (10,000 × g for 10 min at 4°C).
3. Dissolve the dried DNA + RNA pellets in sterile water. No attempt was made to remove linearized plasmid DNA template, as the subsequent packaging step with TMV CP is selective for ssRNA (Jupin *et al.*, 1989) and only full-length transcripts that contain the 3′-proximal OAS (Fig. 1).

Encapsidation of mRNA by TMV CP in Vitro

1. Prepare virions by the method of Leberman (1966) and remove the TMV CP from 0.1–1.0 g samples of virions using high-pH dialysis as described by Durham (1972). CP was stored as helical aggregates in sodium acetate buffer, $I = 0.1$, pH 5.0, at 4°C for up to 6 months.
2. Prepare assembly competent 20S oligomers of CP by transferring samples of helical aggregates to sodium phosphate buffer, $I = 0.1$, pH 7.0, for at least 12 h at 4°C followed by the same buffer at 20°C for at least 18 h.
3. Perform OAS + GUS mRNA packaging reactions under TMV assembly conditions (Durham, 1972) in the presence of 1 m*M* EDTA (sodium salt), pH 7.0, to remove inhibitory divalent cations. CP : GUS mRNA ratios were estimated to be between 50 : 1 and 100 : 1.
4. After packaging, chill the reactions on ice to disrupt the 20S CP aggregates, dilute with water, layer over a 1-cm pad of 30% (w/v) sucrose in 1 m*M* EDTA, pH 7.5, and centrifuge at 36,000 rpm for 18 h at 4°C in a Beckman type 40 rotor.
5. Suspend the pellets of encapsidated GUS mRNA gently in sterile water at about 1 μg/μl, made 1–2 m*M* in $CaCl_2$ and treat with micrococcal nuclease (Boehringer) at 50–300 units/ml for 30 min

at 20°C. Digestion of any unencapsidated or partially packaged GUS mRNA was then stopped by adding 5–10 m*M* EGTA (sodium salt), pH 7.5.

Preparation, Electroporation, and Processing of Tobacco Protoplasts

1. Prepare samples of 1×10^6 protoplasts by the method of Motoyoshi *et al.* (1973) from peeled leaves of control or CP^+ transgenic tobacco [*Nicotiana tabacum* cv. Xanthi; CP^+ line 3404 (Powell Abel *et al.*, 1986)].
2. Suspend protoplasts in 1 ml 0.7 *M* mannitol and electroporate with a single 200-ms square-wave pulse at 200 V/cm (Watts *et al.*, 1987) in the presence of 50 μg/ml GUS pseudovirus particles, or 2.5 μg/ml unencapsidated GUS mRNA [TMV or pseudovirions = 5% (w/w) RNA].
3. After electroporation, incubate protoplasts on ice for 10 min, warm to 20°C for 10 min, wash once by low-speed centrifugation to remove excess inoculum, and resuspend in 10 ml culture medium (Motoyoshi *et al.*, 1973).
4. Incubate protoplasts with continuous low illumination (~20 $\mu E/m^2/s$) at 25°C for 18 h before assaying for GUS activity.
5. To assay for GUS activity, gently centrifuge protoplasts and suspend each pellet (1×10^6 protoplasts) in 0.5 ml extraction buffer (Jefferson, 1987). Lyse the protoplasts by 15 passages through a 26-gauge syringe needle.
6. Collect supernatants by microcentrifugation (12,000 × g for 10 min at 4°C). Measure total soluble protein by the method of Bradford (1976) in 10-μl aliquots of supernatant. Assay GUS activity spectrofluorometrically using 100 μg total soluble protein and MUG as substrate (Jefferson, 1987; Wilson *et al.*, Chapter 1; Gallagher, Chapter 3; Naleway, Chapter 4; Martin *et al.*, Chapter 2).

Discussion

The GUS assay system has been of immense value in comparing the early structural events of pseudovirus disassembly and reporter gene expression in control and "protected" CP^+ transgenic tobacco plants and protoplasts. We anticipate that GUS will continue to provide a useful reporter system for studying plant virus–cell interactions in general. The ease of manipulating *in vitro* transcription plasmids to test various 5′- and/or 3′-RNA sequences for their effects on GUS mRNA expression per se or on the stability of the resultant pseudovirus particles is already clear to us.

One novel means by which production of GUS pseudovirus particles could be increased is apparent from experiments performed some years ago (Sleat *et al.*, 1988a) with stably transformed tobacco plants which expressed Ω′-CAT-OAS-*nos* $poly(A)^+$ chimeric RNAs from a CaMV 35S promoter. Following a 2- to 3-week systemic infection of these tobaccos by TMV, large amounts (several hundreds of milligrams) of ~60-nm-long CAT–pseudovirus particles could be extracted and fractionated from true virions (300 nm long). One secondary effect of transcapsidation of the CAT mRNAs by TMV CP produced in huge amounts during virus infection was that the CAT activity of leaf extracts was measurably reduced (Sleat *et al.*, 1988b). It will be interesting to use the more quantifiable GUS assay system to perform analogous experiments and to extend such complementation studies in virus-infected plant tissues to tell us more about the molecular interactions that occur and their compartmentation.

Acknowledgments

J. K. Osbourn was supported by a Research Studentship from the John Innes Foundation. This work was also supported in part by Diatech Ltd., London. We thank Kitty Plaskitt for the electron micrograph shown in Fig. 2.

References

Ahlquist, P. (1986). *In vitro* transcription of infectious viral RNA from cloned cDNA. *Methods Enzymol.* 118:704–716.

Bradford, M. M. (1976). A rapid and sensitive method for the quantitation of microgram quantities of protein utilizing the principle of protein-dye binding. *Anal. Biochem.* 72:248–254.

Cuozzo, M., O'Connell, K. M., Kaniewski, W., Fang, R-X., Chua, N-H., and Tumer, N. E.. (1988). Viral protection in transgenic tobacco plants expressing the cucumber mosaic virus coat protein or its antisense RNA. *Bio/Technology* 6:549–557.

Durham, A. C. H. (1972). Structure and roles of the polymorphic forms of tobacco mosaic virus protein. I. Sedimentation studies. *J. Mol. Biol.* 67:289–305.

French, R., Janda M., and Ahlquist, P. (1986). Bacterial gene inserted in an engineered RNA virus: Efficient expression in monocotyledonous plant cells. *Science* 231:1294–1297.

Gallie, D. R., and Kado, C. I. (1989). A translational enhancer derived from tobacco mosaic virus is functionally equivalent to a Shine-Dalgarno sequence. *Proc. Natl. Acad. Sci. USA* 86:129–132.

Gallie, D. R., Sleat, D. E., Watts, J. W., Turner, P. C., and Wilson, T. M. A. (1987a). A comparison of eukaryotic viral 5′-leader sequences as enhancers of mRNA expression *in vivo*. *Nucleic Acids Res.* 15:8693–8711.

Gallie, D. R., Sleat, D. E., Watts, J. W., Turner, P. C., and Wilson, T. M. A. (1987b). The 5′-leader sequence of tobacco mosaic virus RNA enhances the expression of foreign gene transcripts *in vitro* and *in vivo*. *Nucleic Acids Res.* 15:3257–3273.

Gallie, D. R., Sleat, D. E., Watts, J. W., Turner, P. C., and Wilson, T. M. A. (1987c). *In vivo* uncoating and efficient expression of foreign mRNAs packaged in TMV-like particles. *Science* 236:1122–1124.

Gallie, D. R., Plaskitt, K. A., and Wilson, T. M. A. (1987d). The effect of multiple dispersed copies of the origin-of-assembly sequence from TMV RNA on the morphology of pseudovirus particles assembled *in vitro*. *Virology* 158:473–476.

Gallie, D. R., Lucas, W. J., and Walbot, V. (1989). Visualizing mRNA expression in plant protoplasts: factors influencing efficient mRNA uptake and translation. *Plant Cell* 1:301–311.

Hayes, R. J., Petty, I. T. D., Coutts, R. H. A., and Buck, K. W. (1988). Gene amplification and expression in plants by a replicating geminivirus vector. *Nature (Lond.)* 334:179–182.

Hemenway, C., Fang, R-X., Kaniewski, W. K., Chua, N-H., and Tumer, N. E. (1988). Analysis of the mechanism of protection in transgenic plants expressing the potato virus X coat protein or its antisense RNA. *EMBO J.* 7:1273–1280.

Hoekema, A., Huisman, M. J., Molendijk, L., van den Elzen, P. J. M., and Cornelissen, B. J. C. (1989). The genetic engineering of two commercial potato cultivars for resistance to potato virus X. *Bio/Technology* 7: 273–278.

Jefferson, R. A. (1987). Assaying chimeric genes in plants: The GUS gene fusion system. *Plant Mol. Biol. Rep.* 5:387–405.

Jupin, I., Sleat, D. E., Watkins, P. A. C., and Wilson, T. M. A. (1989). Direct recovery of *in vitro* transcripts in a form suitable for prolonged storage and shipment at ambient temperatures. *Nucleic Acids Res.* 17:815.

Kozak, M. 1989. The scanning model for translation: an update. *J. Cell Biol.* 108:229–241.

Lawson, C., Kaniewski, W., Haley, L., Rozman, R., Newell, C., Sanders, P., and Tumer, N. (1990). Engineering resistance to mixed virus infection in a commercial potato cultivar: resistance to potato virus X and potato virus Y in transgenic Russet Burbank. *Bio/Technology* 8:127–134.

Leberman, R. (1966). The isolation of plant viruses by means of "simple" coacervates. *Virology* 30:341–347.

Loesch-Fries, L.S., Merlo, D., Zinnen, T., Burhop, L., Hill, K., Krahn, K., Jarvis, N., Nelson, S., and Halk, E. (1987). Expression of alfalfa mosaic virus RNA 4 in transgenic plants confers virus resistance. *EMBO J.* 6: 1845–1851.

Melton, D.A., Krieg, P. A., Rebagliati, M. R., Maniatis, T., Zinn, K., and Green, M. R. (1984). Efficient *in vitro* synthesis of biologically active RNA and RNA hybridization probes from plasmids containing a bacteriophage SP6 promoter. *Nucleic Acids Res.* 12:7035–7056.

Motoyoshi, F., Bancroft, J. B., Watts, J. W., and Burgess, J. (1973). The infection of tobacco protoplasts with cowpea chlorotic mottle virus and its RNA. *J. Gen. Virol.* 20:177–193.

Mundry, K. W., Watkins, P. A. C., Ashfield, T., Plaskitt, K. A., Eisele-Walter, S., and Wilson, T. M. A. (1991). Complete uncoating of the 5′-leader sequence of tobacco mosaic virus RNA occurs rapidly and is required to initiate cotranslational virus disassembly *in vitro*. *J. Gen. Virol.*, 72:769–777.

Nejidat, A., and Beachy, R. N. (1989). Decreased levels of TMV coat protein in transgenic tobacco plants at elevated temperatures reduce resistance to TMV infection. *Virology* 173:531–538.

Nejidat, A., and Beachy, R. N. (1990). Transgenic tobacco plants expressing a coat protein gene of tobacco mosaic virus are resistant to some other tobamoviruses. *Mol. Plant-Microbe Interac.* 3:247–251.

Nelson, R. S., Powell Abel, P., and Beachy, R. N. (1987). Lesions and virus accumulation in inoculated transgenic tobacco plants expressing the coat protein gene of tobacco mosaic virus. *Virology* 158:126–132.

Osbourn, J. K., Watts, J. W., Beachy, R. N., and Wilson, T. M. A. (1989). Evidence that nucleocapsid disassembly and a later step in virus replication are inhibited in transgenic tobacco protoplasts expressing TMV coat protein. *Virology* 172:370–373.

Plaskitt, K. A., Watkins, P. A. C., Sleat, D. E., Gallie, D. R., Shaw, J. G., and Wilson, T. M. A. (1988). Immunogold labeling locates the site of disassembly and transient gene expression of tobacco mosaic virus-like pseudovirus particles *in vivo*. *Mol. Plant-Microbe Interact.* 1:10–16.

Powell Abel, P., Nelson, R. S., De, B., Hoffmann, N., Rogers, S. G., Fraley, R. T., and Beachy, R. N. (1986). Delay of disease development in transgenic plants that express the tobacco mosaic virus coat protein gene. *Science* 232:738–743.

Register, J. C., and Beachy, R. N. (1988). Resistance to TMV in transgenic plants results from interference with an early event in infection. *Virology* 166:524–532.

Register, J. C., and Beachy, R. N. (1989). Effect of protein aggregation state on coat protein-mediated protection against tobacco mosaic virus using a transient protoplast assay. *Virology* 173:656–663.

Sanford, J. C. (1988). The biolistic process. *TIBTECH* 6:299–302.

Shaw, J. G., Plaskitt, K. A., and Wilson, T. M. A. (1986). Evidence that tobacco mosaic virus particles disassemble cotranslationally *in vivo*. *Virology* 148:326–336.

Sleat, D. E. (1987). Packaging and expression of recombinant RNAs. Ph.D. Thesis, University of Liverpool, U.K., Chapter 6.

Sleat, D.E., Turner, P. C., Finch, J. T., Butler, P. J. G., and Wilson, T. M. A. (1986). Packaging of recombinant RNA molecules into pseudovirus particles directed by the origin-of-assembly sequence of tobacco mosaic virus RNA. *Virology* 155:299–308.

Sleat, D. E., Gallie, D. R., Jefferson, R. A., Bevan, M. W., Turner, P. C., and Wilson, T. M. A. (1987). Characterisation of the 5'-leader of tobacco mosaic virus RNA as a general enhancer of translation *in vitro*. *Gene* 60:217–225.

Sleat, D. E., Gallie, D. R., Watts, J. W., Deom, C. M., Turner, P. C., Beachy, R. N., and Wilson, T. M. A. (1988a). Selective recovery of foreign gene transcripts as virus-like particles in TMV-infected transgenic tobaccos. *Nucleic Acids Res.* 16:3127–3140.

Sleat, D. E., Plaskitt, K. A., and Wilson, T. M. A. (1988b). Selective encapsidation of CAT gene transcripts in TMV-infected transgenic tobacco inhibits CAT synthesis. *Virology* 165:609–612.

Stark, D. M., and Beachy, R. N. (1989). Protection against potyvirus infection in transgenic plants: evidence for broad spectrum resistance. *Bio/Technology* 7:1257–1262.

Takamatsu, N., Ishikawa, M., Meshi. T., and Okada, Y. (1987). Expression of bacterial chloramphenicol acetyltransferase gene in tobacco plants mediated by TMV-RNA. *EMBO J.* 6:307–311.

Tumer, N. E., O'Connell, K. M., Nelson, R. S., Sanders, P. R., Beachy, R. N., Fraley, R. T., and Shah, D. M. (1987). Expression of alfalfa mosaic virus coat protein gene confers cross protection in transgenic tobacco and tomato plants. *EMBO J.* 6:1181–1188.

van Dun, C. M. P., and Bol, J. F. (1988). Transgenic tobacco plants accumulating tobacco rattle virus coat protein resist infection with tobacco rattle virus and pea early browning virus. *Virology* 167:649–652.

van Dun, C. M. P., Overduin, B., van Vloten-Doting, L., and Bol, J. F. (1988). Transgenic tobacco expressing tobacco streak virus or mutated alfalfa mosaic virus coat protein does not protect against alfalfa mosaic virus infection. *Virology* 164:383–389.

Ward, A., Etessami, P., and Stanley, J. (1988). Expression of a bacterial gene in plants mediated by infectious geminivirus DNA. *EMBO J.* 7:1583–1587.

Watts, J. W., King, J. M., and Stacey, N. J. (1987). Inoculation of protoplasts with viruses by electroporation. *Virology* 157:40–46.

Wilson, T. M. A. (1984). Cotranslational disassembly of tobacco mosaic virus *in vitro*. *Virology* 137:255–265.

Wilson, T. M. A., and Watkins, P. A. C. (1986). Influence of exogenous viral coat protein on the cotranslational disassembly of tobacco mosaic virus (TMV) particles *in vitro*. *Virology* 149:132–136.

Wilson, T. M. A., Saunders, K., Dowson-Day, M. J., Sleat, D. E., Trachsel, H., and Mundry, K. W. (1990). Effects of the 5′-leader sequence of tobacco mosaic virus RNA, or derivatives thereof, on foreign mRNA and native viral gene expression. In "NATO ASI Series: Post-transcriptional Control of Gene Expression" (eds. J. E. G. McCarthy and M. F. Tuite). Vol. H49, pp. 261–275. Springer-Verlag, Berlin.

Wu, A.-Z., Dai, R-M., Shen, X-R., and Sun, Y. K. (1984). Abstracts of the Sixth International Congress of Virology, Sendai, Japan, p. 231.

11 Sensitive Assay for the Excision of Transposable Elements Using β-Glucuronidase Reporter Gene

E. Jean Finnegan
CSIRO
Division of Plant Industry
Canberra, Australia

Transposable elements, such as the maize element *Activator* (*Ac*), are able to move from one location in the plant genome to another. There is no obvious phenotype that can be directly associated with transposable elements, unless the element has inserted into a gene that conditions a visible phenotype, disrupting expression of that gene. The maize kernel provides a fine example for studying the movement of transposable elements because genes involved in both starch and storage protein biosynthesis as well as genes affecting biosynthesis of the anthocyanin pigments of the aleurone layer of the kernel act as natural reporter genes for studying transposon movement. Insertion of a transposable element within these genes can inactivate the gene, dramatically altering the phenotype of the kernel without affecting its viability. The frequency and timing of excision can be investigated by scoring the number and size of sectors in which gene activity has been restored by excision of the element (McClintock, 1951, 1956; Brink and Nilan, 1952). Genes affecting flower pigmentation of *Antirrhinum majus* have been used in a similar way to monitor the activity of *Tam* transposable elements (Bonas *et al.*, 1984; Martin *et al.*, 1985; Sommer *et al.*, 1985; Upadhyaya *et al.*, 1985; Coen *et al.*, 1986).

GUS Protocols: Using the GUS Gene as a Reporter of Gene Expression

Transposable elements from maize and *Antirrhinum* have been cloned and introduced into heterologous plant hosts that do not have well-characterized transposable element systems (Baker *et al.*, 1986; Van Sluys *et al.*, 1987; Knapp *et al.*, 1988; Yoder *et al.*, 1988). This opens the way to examine the functions encoded on a transposon by reverse genetics, where a modified transposable element can be introduced into a new host by *Agrobacterium*-mediated transformation. Plant transposable elements have been used to clone genes, the products of which are unknown, in their natural host (Bingham *et al.*, 1981; Fedoroff *et al.*, 1981; Martin *et al.*, 1985; Moerman *et al.*, 1986), and the potential for an introduced element to act as an insertional mutagen and thus for "gene tagging" in a foreign host has been recognized.

Movement of the introduced transposable element can be followed by Southern analysis; excision of the element results in the appearance of a band of new mobility, which hybridizes to DNA sequences that were adjacent to the element in the introduced T-DNA, corresponding to an "empty site." These analyses are time-consuming, require substantial amounts of plant material, and provide only limited information. For example, while the appearance of an excision band indicates that the element is mobile, the intensity of this band does not distinguish between the contribution from a single excision event that gave rise to a large sector of cells and that from a number of small sectors each arising from a different excision event. In contrast, a visual assay of excision allows both the size and number of sectors corresponding to individual excision events to be recorded. In addition, it is not possible to determine in which tissues excision has occurred by a Southern analysis of total plant DNA.

A phenotypic assay for excision of the maize transposable element *Activator* (*Ac*) in a foreign host was developed by Baker *et al.* (1987). *Activator* was cloned into the untranslated leader of the neomycin phosphotransferase gene (NPT II), disrupting gene expression. Appearance of kanamycin resistant callus was associated, by Southern analysis, with excision of *Ac* and restoration of the NPT II gene (Baker *et al.*, 1987). Selection was applied immediately following transformation or after a period of growth on nonselective medium. Although this phenotypic assay enabled detection of *Ac* excision without the need for Southern analysis, it did not allow either the timing or frequency of excision to be followed at the whole-plant level.

A visual assay for *Ac* excision was developed by Jones *et al.* (1989) using the bacterial streptomycin phosphotransferase gene (SPT) as a reporter. Excision of *Ac* from the untranslated leader restores gene expression, resulting in green (streptomycin resistant) sectors on a bleached (sensitive) background in cotyledons of tobacco seedlings germinated on medium containing streptomycin. The size and frequency of the green sectors gives an indication of the timing and frequency of excision events. The cloning of a chimeric β-glucuronidase gene (GUS) for expression in plants provided the basis for developing a new sensitive assay for the excision of transposable elements in a foreign host. *Activator* was cloned into the untranslated leader of the chimeric 35S-GUS-nos3′ gene, inactivating the gene. Excision of *Ac*, which restores the gene, can be followed indirectly by assaying for GUS expression, either by a fluorometric assay or by histochemical staining.

Excision of *Ac* was monitored in a number of independent transgenic plants by assaying for GUS activity (Table 1 and Color Plate 6). Southern analysis of DNA isolated from the same plants confirmed that, for most plants, GUS activity correlated with the appearance of an excision band (Figure 1). No excision band was detected in DNA isolated from plant p*Ac* G-1, although histochemical staining of leaf tissue taken from this plant showed small sectors expressing the GUS gene, indicating that excision of *Ac* occurred late in development. The proportion of cells in which excision had occurred, as judged by histochemical staining, was too low (<10% total cell population) to yield an excision band that could be detected by Southern hybridization.

The sensitivity of both the fluorometric and histochemical assays for GUS activity, together with the speed of the assays and the small amount of plant tissue sacrificed, makes the GUS reporter gene an attractive choice to monitor excision of an introduced transposable element in transgenic plants.

Table 1

GUS Activity in Extracts Made from Individual Leaves Taken from Untransformed *N. tabacum* or from Plants Transformed with pBI121, pΔ*Ac* G, or p*Ac* G[a]

Plant	Leaf 1	Leaf 2
Untransformed	28	—
pBI121	1500	1585
pΔ*Ac* G	59	—
p*Ac* G-1	319	105
p*Ac* G-3	498	324
p*Ac* G-4	1625	1665

[a] Values are the average of assays done in duplicate using an equal amount of protein per assay and are expressed as fluorimeter units.

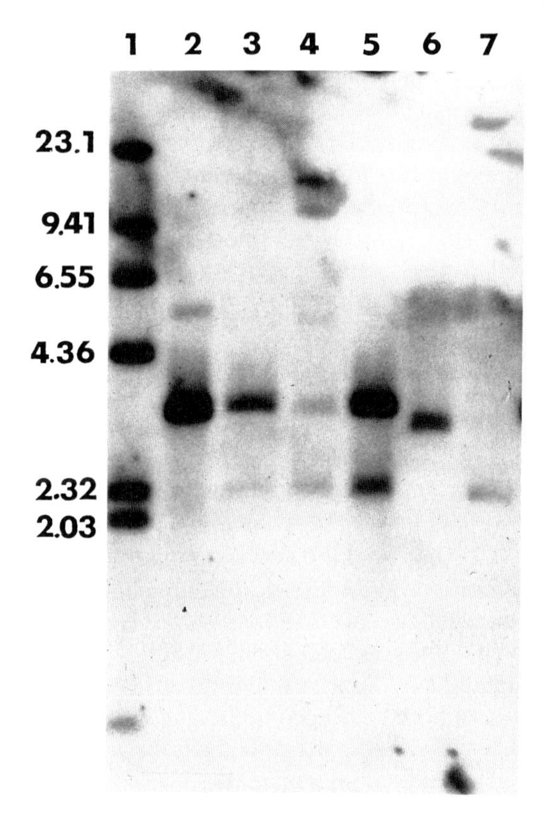

Fig. 1 Southern analysis of DNA extracted from transgenic tobacco plants. DNA was digested with XbaI plus EcoRI and was probed with a fragment containing the GUS coding region. This probe will hybridize to a 3.3-kb or 3.1-kb fragment if the 35S-GUS-nos3′ gene is interrupted by *Ac* or Δ*Ac*, respectively. If excision occurs, restoring the GUS chimaeric gene, the probe will hybridize to a 2.3-kb fragment. Constructs p*Ac* G, pΔ*Ac* G, and pBI121 are described in Figure 2. An excision band can be seen in DNA from plants p*Ac* G-2, -3, and -4 (lanes 3, 4, and 5, respectively). The probe also hybridizes to the interrupted gene in DNA from p*Ac* G-1, -2, -3, and -4 and pΔ*Ac* G (lanes 2–6, respectively). The probe hybridizes to a 2.3-kb band in DNA isolated from a plant transformed with pBI121 (lane 7).

Protocols

Transposable Element–Reporter Gene Constucts

When cloning a transposable element into the untranslated leader of the 35S-GUS-nos3′ reporter gene, two points should be taken into consideration. The element should be closely flanked by suitable restriction sites so that large stretches of exogenous DNA (flanking the element) are not left in the leader after excision. Flanking DNA should be scanned for ATG codons that may result in erroneous translation initiation.

The *Ac* element in the construct described in this chapter was flanked by 21 and 26 bp. It was cloned as a BglII fragment into the unique BamHI site in the plasmid pBI121 (Jefferson *et al.*, 1987), which lies in the untranslated leader of the 35S-GUS-nos3′ gene. The constructs are shown in Figure 2.

Transformation

The transposable element–reporter gene construct can be introduced into the host plant of choice by *Agrobacterium*-mediated transformation or by direct DNA transfer. In this example, plasmids pBI121, p*Ac* G, and pΔ*Ac* G were mobilized into *Agrobacterium tumefaciens* LBA4404 (Hoekema *et al.*, 1983) by triparental mating (Ditta *et al.*, 1980). Transformation of *Nicotiana tabacum* cv W38 was as described in Ellis *et al.* (1987).

Assay for Excision of the Transposable Element

Transgenic plants should be maintained in sterile culture to avoid bacterial or fungal contamination, which could contribute to the GUS activity recorded in either fluorometric or histochemical assays.

Histochemical Staining

1. Take leaves, or other tissue to be assayed, from transgenic plants and cut small sections by hand. To cut thin sections of leaf lamina, hold two razor blades parallel and slice across the leaf, supported on a wax block (for further details on hand sectioning see Craig, Chapter 8; Martin *et al.*, Chapter 2; Stomp, Chapter 7). Although tissues can be sectioned as required for microscopy/photography after staining for GUS activity, the cutting of small sections prior to staining facilitates both the entry of fixative and GUS substrate and the removal of chlorophyll after staining. Tissue can be fixed lightly using glutaraldehyde, as described

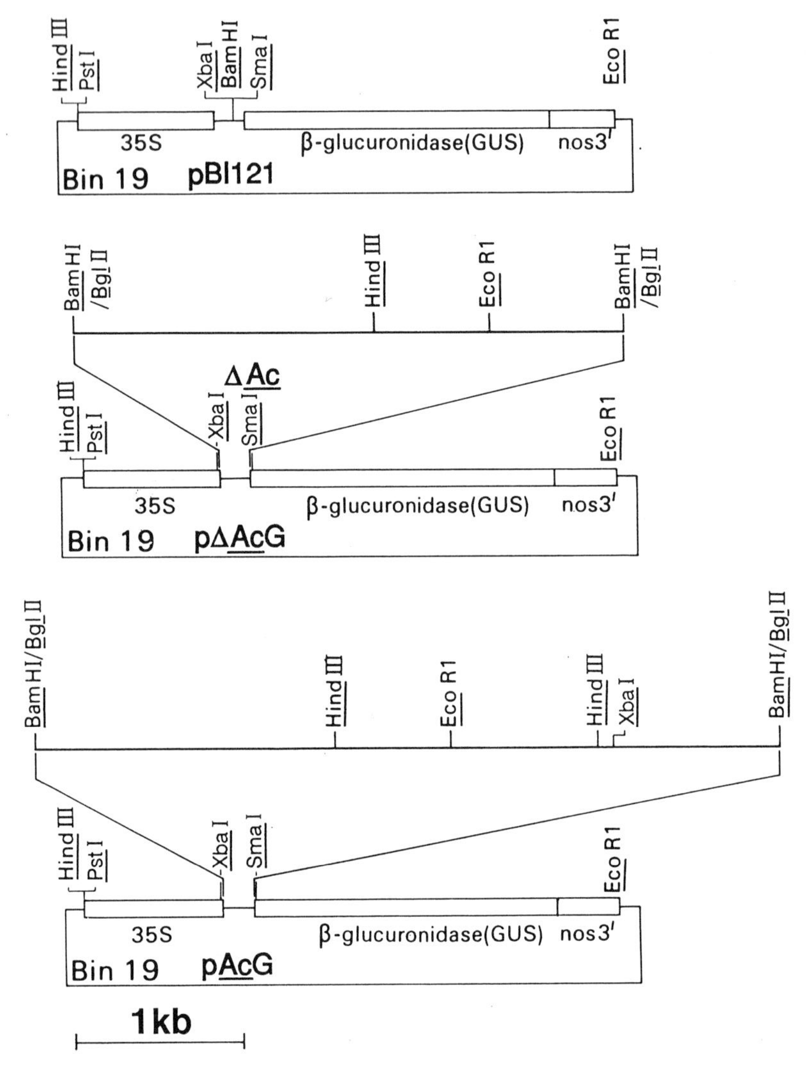

Fig. 2 Restriction maps for the binary vectors pBI121, pΔ*Ac* G, and p*Ac* G. Abbreviations: 35S, cauliflower mosaic virus 35S promoter; GUS, coding region of *E. coli* β-glucuroidase gene; nos3′, 3′ termination region from nopaline synthase; Bin 19, binary vector (Bevan, 1984).

below, or can be stained without prior fixing. As fixation facilitates penetration of GUS substrate, unfixed tissue should be stained with GUS assay buffer containing 10% DMSO (dimethyl sulphoxide), which also facilitates substrate penetration.

2. Fix tissue by immersion of the hand-cut sections in ice-cold 1% glutaraldehyde in 25 m*M* sodium phosphate, pH 7.0, in glass vials. Infiltrate the fixative into the tissue by brief evacuation, using a water vacuum, and repeat several times to ensure good penetration. Cap the vials and fix tissues on ice for a total of 30 min.
3. Wash tissue (4 × 5 min) in ice-cold 25 m*M* sodium phosphate.
4. To stain for GUS activity, replace the buffer with GUS assay buffer [0.1 *M* NaH_2PO_4, pH 7.0, 0.5 m*M* potassium ferricyanide, 0.5 m*M* potassium ferrocyanide, 1 m*M* 5-bromo-4-chloro-3-indoyl β-D-glucuronide (X-glucuronide)] and incubate the tissue at 37°C in the dark for approximately 16 h. At this stage blue-stained sectors are usually visible, but resolution can be improved by removal of chorophyll pigment.
5. To remove chlorophyll, immerse the tissue in 25% ethanol and agitate gently several times over a 15-min period; remove the ethanol and replace with 50% ethanol. Repeat this procedure progressing through a graded series of ethanol washes (25, 50, 75, 90, 95, 100%). Chlorophyll is visibly extracted by ethanol at ≥90%, but repeated washing in absolute ethanol may be necessary to remove chlorophyll completely. This dehydrates the tissue, making it possible to infiltrate it with immersion oil, which has the appropriate refractive index, to facilitate photography of stained tissue sections.
6. Replace the ethanol with acetone (5 × 10 min washes), then soak the tissue in a 1 : 1 mix of acetone/immersion oil for at least 1 h and finally in undiluted immersion oil for at least 1 h. Remove all traces of acetone under vacuum; leaf pieces should not be left in acetone because extraction of GUS reaction product may occur. Tissue can be stored in immersion oil for extended periods without changing the pattern or intensity of staining.
7. For photography, mount the tissue under a coverslip and photograph under a stereomicroscope using Ektachrome professional 50 ASA or 160 ASA film.

Fluorometric Assay

1. Using a pestle and mortar, grind individual leaves (2–3 cm) taken from transgenic plants in 1 ml 0.25 *M* Tris-HCl, pH 7.5, then transfer to an eppendorf tube and centrifuge at 4°C for 5 min to pellet the debris.

2. Carefully remove the supernatant, place in a clean Eppendorf tube, and store on ice.
3. Determine the protein content of each extract using the Coomassie blue method of Bradford (1976) and adjust the volume of extract with 0.25 *M* Tris HCl, pH 7.5, so that the protein content is identical for each extract.
4. Mix equal volumes (50 μl) of extract and assay buffer (50 m*M* NaH_2PO_4, pH 7.0, 10 m*M* EDTA, 10 m*M* β-mercaptoethanol, 1 m*M* 4-methylumbelliferyl glucuronide) and incubate at 37°C for 30 min. Include a blank sample using 50 μl 0.25 *M* Tris HCl, pH 7.5, plus 50 μl of assay buffer, in the assay to provide a zero reading.
5. Stop the reaction by adding 1.5 ml 0.2 *M* Na_2CO_3 and measure the fluorescence in a fluorimeter, with excitation at 365 nm and emission at 455 nm.

Southern Analysis

DNA extraction, DNA digestion, gel electrophoresis, transfer of DNA, and hybridization can be done according to any published method (for example, Sambrook *et al.,* 1989). DNA used in the analysis described here was isolated using the method described in Taylor and Powell (1982). To detect excision bands DNA should be cut with restriction enzyme(s) that generate a fragment, the size of which will change if the transposable element excises. The probe used must be homologous to DNA flanking the element rather than to the element itself, that is, the probe could be derived from the 35S promotor, the GUS coding region, the nos3′, or could include the 35S-GUS junction disrupted by the cloning of the element. Suitable enzyme/probe combinations for the detection of an excision band for 35S-GUS-nos3′ with an element cloned into the BamHI site include EcoRI plus XbaI, probe GUS or nos3′; PstI plus SmaI, probe 35S; EcoRI plus HindIII, probe GUS, nos3′, or 35S. It is acceptable to use an enzyme that has a recognition site within the cloned element providing the fragment generated, which hybridizes to the probe, differs in size from the predicted excision fragment.

Discussion

Several phenotypic assays for excision of transposable elements in a heterologous host have been described (Baker *et al.*, 1987; Jones *et al.*,

1989; Masson and Fedoroff, 1989; Finnegan *et al.*, 1989). The *Ac*–GUS reporter gene combination described here provides an excellent means to study transposable element excision. It is not only more sensitive than a conventional Southern analysis but is also much quicker, requires only a small amount of tissue, and provides information about the timing and frequency of excision. Although the assay is destructive (unlike the SPT-based visual assay), it is more general because excision can be followed in most plant tissue. Localization of excision to particular cells is possible because of the cell-autonomous expression of the GUS gene and the insoluble nature of the GUS reaction product. While the assays based on the NPT II and SPT reporter genes do have some attractive features, both are of limited applicability when compared to that using the GUS reporter gene.

Cells in which excision has occurred can be selected using an *Ac* : NPT II reporter system (Baker *et al.*, 1987). Although it is possible to estimate the frequency of *Ac* excision in individual protoplasts or in callus, by comparing the number of kanamycin-resistant cells to the total population, it is difficult to obtain a measure of *Ac* excision for the whole plant. In addition, if kanamycin selection is applied to isolate transgenic plants in which excision has occurred, it cannot be used to monitor excision in subsequent generations.

The visual assay using an *Ac* : SPT reporter system facilitated an elegant analysis of *Ac* excision in progeny of transgenic plants (Jones *et al.*, 1989). Germination of seeds on medium containing streptomycin provided information on the frequency (the number of green sectors) and timing (sectored or fully green cotyledons) of excision. The assay is nondestructive, as seedlings can be recovered for further analysis by transferring to streptomycin-free medium. This visual assay is limited to an analysis of *Ac* excision in chlorophyll-producing tissue in plants that are susceptible to streptomycin. It may be difficult to monitor *Ac* excision in mature plants because of the time required to bleach already green plant tissue cultured on medium containing steptomycin.

The phenotypic assay based on the GUS reporter gene has some advantages when compared to the SPT-based assay. Detection of *Ac* is not limited to green tissue; excision can be monitored in any tissue in which the 35S promoter is active by either fluorometric or histochemical assay. Thus it is possible to follow excision of a transposable element in different tissues throughout plant development. The timing and frequency of excision can be estimated by examining the size and number of stained sectors seen in the histochemical assay. In addition, because each sector arises from a single cell in which the element has excised, the size and shape of any sector may lead to an understanding of plant development. It should be noted however, that histolocalization is not without potential problems, as Benfey and Chua (1989) have observed variation in the tissue-specific expression of a 35S GUS gene in a number of independent transgenic petunia and tobacco. Although the *Ac* transposable element has been used to develop this assay, it can be readily adapted for other transposable elements (e.g., *Spm;* Masson and Fedoroff, 1989) and can be used in any plant that is amenable to transformation.

References

Baker, B., Schell, J., Lörz, H., and Fedoroff, N. (1986). Transposition of the maize controlling element "*Activator*" in tobacco. *Proc. Natl. Acad. Sci. USA* 83, 4844–4848.

Baker, B., Coupland, G., Fedoroff, N., Starlinger, P., and Schell, J. (1987). Phenotypic assay for excision of the maize controlling element *Ac* in tobacco. *EMBO J.* 6, 1547–1554.

Benfey, P. N., and Chua, N.-H. (1989). Regulated genes in transgenic plants. *Science* 244, 174–181.

Bevan, M. (1984). Binary *Agrobacterium* vectors for plant transformation. *Nucleic Acids Res.* 12, 8711–8721.

Bingham, P. M., Levis, R., and Rubin, G. M. (1981). Cloning of DNA sequences from the *white* locus of *D. melanogaster* by a novel and general method. *Cell* 25, 693–704.

Bonas, U., Sommer, H., and Saedler, H. (1984). The 17-kb *Tam*l element of *Antirrhinum majus* induces a 3 bp duplication upon integration into the chalcone synthase gene. *EMBO J.* 3, 1015–1019.

Bradford, M. M. (1976). A rapid and sensitive method for the quantitation of microgram quantities of protein utilizing the principle of protein-dye binding. *Anal. Biochem.* 72, 248–254.

Brink, R. A., and Nilan, R. A. (1952). The relation between light variegated and medium variegated pericarp in maize. *Genetics* 37, 519–544.

Coen, E. S., Carpenter, R., and Martin, C. (1986). Transposable elements generate novel spatial patterns of gene expression in *Antirrhinum majus. Cell* 47, 285–286.

Ditta, G., Stanfield, S., Corbin, D., and Helinski, D. R. (1980). Broad host range DNA cloning system for Gram negative bacteria: construction of a gene bank of *Rhizobium meliloti. Proc. Natl. Acad. Sci. USA* 77, 7347–7351.

Ellis, J. G., Llewellyn, D. J., Dennis, E. S., and Peacock, W. J. (1987). Maize *Adh*1 promoter sequences control anaerobic regulation: Addition of upstream promoter elements for constitutive genes is necessary for expression in tobacco. *EMBO J.* 6, 11–16.

Fedoroff, N. V., Furtek, D. B., and Nelson, O. E., Jr. (1984). Cloning of the *bronze* locus in maize by a simple and generalizable procedure using the transposable controlling element *Activator* (*Ac*). *Proc. Natl. Acad. Sci. USA* 81, 3825–3829.

Finnegan, E. J., Taylor, B. H., Craig, S., and Dennis, E. S. (1989). Transposable elements can be used to study cell lineages in transgenic plants. *Plant Cell* 1, 757–764.

Hoekema, A., Hirsch, P. R., Hooykaas, P. J. J., and Schilperoort, R. A. (1983). A binary plant vector strategy based on separation of *vir* and T-region of the *Agrobacterium tumefaciens* Ti plasmid. *Nature* 303, 179–180.

Jefferson, R. A., Kavanagh, T. A., and Bevan, M. W. (1987). GUS fusions: *β-Glucuronidase* as a sensitive and versatile gene fusion marker in higher plants. *EMBO J.* 6, 3901–3907.

Jones, J. D., Carland, F. M., Maliga, P., and Dooner, H. K. (1989). Visual detection of transposition of the maize element *Activator* (*Ac*) in tobacco seedlings. *Science* 244, 204–207.

Knapp, S., Coupland, G., Uhrig, H., Starlinger, P., and Salamini, F. (1988). Transposition of the maize transposable element *Ac* in *Solanum tuberosum*. *Mol. Gen. Genet.* 213, 285–290.

Martin, C., Carpenter, R., Sommer, H., Saedler, H., and Coen, E. S. (1985). Molecular analysis of instability in flower pigmentation of *Antirrhinum majus* following isolation of *pallida* locus by transposon tagging. *EMBO J.* 4, 1625–1630.

Masson, P., and Fedoroff, N. V. (1989). Mobility of the maize suppressor-mutator element in transgenic tobacco cells. *Proc. Natl. Acad. Sci. USA* **86,** 2219–2223.

McClintock, B. (1951). Chromosome organization and gene expression. *CSH Symp. Quant. Biol.* 16, 13–47.

McClintock, B. (1956). Controlling elements and the gene. *CSH Symp. Quant. Biol.* 21, 197–216.

Moerman, D. G., Benian, G. M., and Waterston, R. H. (1986). Molecular cloning of the muscle gene *unc*-22 in *Caenorhabditis elegans* by *Tc*1 transposon tagging. *Proc. Natl. Acad. Sci. USA* 83, 2579–2583.

Sambrook, J., Fritsch, E. F., and Maniatis, T. (1989). "Molecular Cloning. A Laboratory Manual." Cold Spring Harbor Laboratory Press, Cold Spring Harbor, N.Y., 2nd ed.

Sommer, H., Carpenter, R., Harrison, B. J., and Saedler, H. (1985). The transposable element *Tam* 3 of *Antirrhinum majus* generates a novel type of sequence alteration upon excision. *Mol. Gen. Genet.* 199, 225–231.

Taylor, B., and Powell, A. (1982). Isolation of plant DNA and RNA. *Focus* 4, 4–6.

Upadhyaya, K. C., Sommer, H., Krebbers, E., and Saedler, H. (1985). The paramutagenic line *niv*-44 has a 5-kb insert, *Tam* 2 in the chalcone synthase gene of *Antirrhinum majus*. *Mol. Gen. Genet.* 199, 201–207.

Van Sluys, M. A., Tempe, J., and Fedoroff, N. (1987). Studies on the introduction and mobility of the maize *Activator* element in *Arabidopsis thaliana* and *Daucus carotta*. *EMBO*. *J*. 6, 3881–3889.

Yoder, J. I., Palys, J., Alpert, K., and Lassner, M. (1988). *Ac* transposition in transgenic tomato plants. *Mol*. *Gen*. *Genet*. 213, 291–296.

12 Anthocyanin Genes as Visual Markers in Transformed Maize Tissues

Ben Bowen
Department of Biotechnology Research
Pioneer Hi-Bred International, Inc.
Johnston, Iowa

Introduction

Not all plant cells are green. Many cell types in a broad range of plant species accumulate red or purple pigments known as anthocyanins (Brouillard, 1988). The formation of these pigments is controlled both by structural genes, which encode enzymes involved with anthocyanin biosynthesis (Mol *et al.*, 1988), and by regulatory genes, which activate expression of the structural genes at the transcriptional level (Ludwig and Wessler, 1990). Many of these genes have been used as markers in unpigmented cells of the appropriate genotype much in the same way that β-glucuronidase (GUS) has been used in cells that lack endogenous β-glucuronidase activity (Meyer *et al.*, 1987; Klein *et al.*, 1989; Ludwig *et al.*, 1990; Goff *et al.*, 1990; Nash *et al.*, 1990; Menssen *et al.*, 1990).

There are several reasons why anthocyanin markers provide an attractive alternative to GUS. First and foremost, visualization of anthocyanin gene expression is nondestructive and requires no fixation or histochemical substrate (Color Plate 7A–E). Second, the accumulation of anthocyanins is cell-autonomous, whereas the pattern of histochemical staining with GUS is not (Color Plate 7, C and D). This is particularly advantageous when using microprojectile bombardment as a DNA delivery system, because cells transiently expressing these genes can be

GUS Protocols: Using the GUS Gene as a Reporter of Gene Expression

counted easily and unambiguously. Furthermore, when unsuitably harsh bombardment conditions are employed, we generally find that cells that have been damaged by microprojectiles can still express GUS (i.e., they will stain blue and X-Gluc), but they will not accumulate anthocyanin, presumably because the latter process requires an intact vacuole and the coordinate expression of many genes. With anthocyanin markers, one can also follow the fate of transformed cell lineages in growing plant tissues (M. Ross, D. Tomes, and B. Bowen, unpublished data) (Color Plate 7E) and observe how genes are expressed in living plants in ways that would be impractical or impossible using GUS histochemistry (Meyer *et al.,* 1987; van der Krol *et al.,* 1988, 1990; Napoli *et al.,* 1990; Benfey *et al.,* 1990).

Much of the work involving transformation with anthocyanin markers has been performed with maize. In this species, the anthocyanin biosynthetic pathway is regulated by two families of transcriptional activators, members of which I shall refer to as R and C proteins, respectively (Ludwig and Wessler, 1990). In the absence of R and/or C expression, the structural genes encoding enzymes in the anthocyanin biosynthetic pathway are silent and no pigments accumulate. Coexpression of genes encoding R and C proteins is necessary for pigment formation (Coe *et al.,* 1988). Whether R or C proteins are encoded by endogenous genes or an introduced template is immaterial. This means that either R or C or both R and C can be used as visual markers in transformed maize cells, depending on their genotype (Ludwig *et al.,* 1990; Goff *et al.,* 1990). Recently it has been found that pigments can also be induced in other monocots (e.g., other Gramineae, some Liliaceae and Orchidaceae) by maize R and C genes (B. Bowen, unpublished data; J. Wong and T. Klein, personal communication). As in maize, coexpression of both R and C was necessary in each species tested.

Anthocyanins can be induced in virtually any maize cell type into which vectors expressing R and/or C can be delivered without damage by microprojectile bombardment (Ludwig *et al.,* 1990; Goff *et al.,* 1990). In this chapter, I will describe two experimental systems that can be used to optimize parameters affecting DNA delivery by the particle gun and to make comparisons between the performance of different vectors in transient assays. Much of the analysis involves comparisons between dose-response curves, where "dose" corresponds to the amount of plasmid-encoded template precipitated on to a fixed number of microprojectiles, and "response" corresponds to the number of red cells counted in a fixed number of bombardments. This approach can also be used to compare transient levels of enzyme activity directed by

35S :: luciferase or 35S :: GUS vectors analogous to those described for R and C here (i.e., pPHI471 and pPHI665). Using current bombardment procedures in practice at Pioneer, red cells can be detected in maize embroygenic callus samples with 10^{-18} mol of template containing both 35S :: R and 35S :: C per bombardment. The lower limits of sensitivity for analogous luciferase and GUS templates are approximately 10^{-16} and 10^{-15}–10^{-14} mol per bombardment, respectively (B. Drummond, unpublished data).

Protocols

DNA and Particle Preparation

1. Purify plasmid DNA from *Escherichia coli* by alkaline lysis and either CsCl density-gradient centrifugation or polyethylene glycol (PEG) precipitation (Sambrook *et al.*, 1989). DNA that shows no smearing on an alkaline gel (Sambrook *et al.*, 1989) will give the most reliable results.
2. Plasmids are premixed in TE (10 m*M* Tris-HCL, 1m*M* EDTA, pH 8.0) buffer and added to 4.375 mg of particles in a total volume of 10 μl. Various kinds of particles can be used for microprojectile bombardment (Klein *et al.*, 1988; Ludwig *et al.*, 1990). Particles that have worked well at Pioneer are from Engelhard (A1570 Flakeless Gold) or from General Electric Co. (1.8 μm average diameter tungsten), but optimum performance must be determined empirically by the investigator (see, for example, Fig. 5B).
3. Resuspend the particles by brief sonication and add 25 μl of 2.5 *M* $CaCl_2$ and 10 μl of 0.1 *M* spermine (free base) with an additional brief sonication.
4. Let the particles settle for 10 min and then remove 30 μl of supernatant before dispensing aliquots for bombardment.

Each plasmid mixture should contain an equal weight of DNA (usually 10 μg). In many instances, it is also preferable to maintain approximately the same copy number of enhancer sequences in each mixture to avoid differential titration of endogenous transcription factors. This is especially important for comparing responses over a wide range of plasmid doses (see below). For example, if a construct expressing R from the cauliflower mosaic virus 35S promoter is being used at several low doses (e.g., 0.001–0.1 μg per tube), 35S enhancer sequences can be kept at about the same copy number by adding appropriate amounts (i.e., 9.9–9.999 μg) of a 35S :: GUS construct of approximately similar size.

Tissue Preparation

Embryogenic Suspension Cells

Most embryogenic maize suspension cell cultures are derived from genotypes that are typically *c r-r/g b/B-b pl* (Coe *et al.*, 1988). Usually it is necessary to introduce both 35S :: R and 35S :: C to induce the anthocyanin pathway (Goff *et al.*, 1990). Cells are frequently grown in low light at 30°C using Murashige and Skoog (MS) based or similar medium containing 2 mg/liter or more of 2,4-dichlorophenoxyacetic acid.

1. 12–72 hr after subculture, sieve the cells through a 710 μm screen and separate from their culture fluid by filtration (e.g., in a Büchner funnel).
2. Resuspend the cells at a density of 25 μg/ml in fresh culture medium and incubate overnight under normal growing conditions or in the presence of an osmoticum such as 0.25 *M* sorbitol (Ludwig *et al.*, 1990).
3. On the following day, pipet 1-ml aliquots into 60 × 20 mm petri dishes containing two filter discs (Qualitative Grade 363, Baxter Scientific Products) prewetted with 1 ml culture medium. A small section of a disposable 5-ml syringe (sterilized in ethanol) can be used as a template to ensure that each aliquot of cells is centrally positioned and has an approximately uniform surface area. If cells are prone to dislodging during bombardment, remove excess media by filtration at this point.
4. Bombard cells according to published procedures (Klein *et al*, 1988; Ludwig *et al.*, 1990; Goff *et al.*, 1990). A single bombardment with 1 μl of particles prepared as described above is usually sufficient for transient expression studies. However, it is important to replicate each treatment at least three times. Place a nylon or Nitex screen (e.g., 50–150 μm) between the target tissue and the stopping plate if red cell numbers are low.
5. After bombardment, supplement the cells with 1 ml fresh culture medium and then incubate in darkness at 30°C for 2–3 days. Count red cells at this point.

Seeds

Seeds from any line of maize that has a functional complement of the structural genes required for anthocyanin biosynthesis (i.e., *A1 A2 Bz1 Bz2 C2*) can be used, but it is preferable that these genes are not expressed in either the aleurone (as in lines that are *c R/r-g* or *C r-g*) or in germinating seedlings (as in lines that are *R/r-g b/B-b*). I routinely use a W22-derived line that is *A1 A2 Bz1 Bz2 C1 C2 B-b pl r-g :: Stadler,* but many commercial lines of maize can also be used. The constitution of alleles at *C1, R, Pl* and *B* should be known, because this will determine whether light will be required for expression in seedlings (i.e., if they are *pl*), or if plasmids expressing R, C, or both R and C will be necessary for detecting expression in aleurone cells (Coe *et al.,* 1988; Ludwig *et al.,* 1990; Goff *et al.,* 1990).

1. Surface-sterilize seeds (e.g., for 20–30 min in 20% Clorox containing 0.05% sodium dodecyl sulfate and a drop of Tween-20), rinse 4–5 times in sterile distilled water, and then leave to imbibe water overnight in a vessel that is covered but not sealed (e.g., a sterile petri dish).
2. Before the seeds are dissected under aseptic conditions, it may help to consult any standard text describing maize kernel anatomy (e.g., Kiesselbach, 1980). During each manipulation, grasp the seed firmly by a pair of forceps. Using a number 11 scalpel blade, make an incision in the pericarp around the edge of the embryo. Remove and discard the pericarp covering the embryo.
3. Remove the embryo by repeatedly inserting a tapered spatula with a flat end around its edge and gently teasing the scutellum surface away from the underlying endosperm.
4. Place the embryo on MS medium containing 2% sucrose and 0.5% Gel-Rite with the plumule uppermost, preferably side-by-side with and in the same orientation as an embryo from a previously dissected seed. Good media contact is essential for both seedling vigor and synchronous germination.
5. Slice the rest of the kernel in half longitudinally along the axis of the displaced embryo. The remaining pericarp can then be peeled off most easily by pulling at the basal hilar region of each kernel half with a fine pair of forceps. However, it is very important not to damage the nucellar membrane or the underlying aleurone cells during this process. Place kernel halves with the aleurone layer uppermost on the same media used for embryo germination, preferably in such a way that matching halves can be distinguished for subsequent randomization (see below).

6. Following dissection, incubate both embryos and kernel halves under illumination at 30°C until the next day.
7. For optimal results, bombard seedlings 18–24 h and kernel halves up to 30 h after dissection. A single bombardment with 1 μl of particles is sufficient to observe transient expression in either target. Seedlings are prone to damage with more than one bombardment.
8. Incubate both seedlings and kernel halves at 30°C following bombardment, under a regime of 18 h of light and 6 h of darkness for 2–3 days, at which point the number of red cells can be counted.

A randomized complete block design is appropriate for comparing treatments with kernel halves, because there is more variation between unmatched kernel halves than there is between those that are matching. I set this up by using a computer program to generate a random series of numbers for each row of a matrix in which the number of columns corresponds to the number of treatments multiplied by four (the number of half-kernels empirically chosen per replicate) and the number of rows corresponds to the number of replicates (usually 3–4). I next assign a numerical series to a set of matching kernel halves that corresponds to the number of columns in the matrix. For each treatment, clusters of four half-kernels (arranged in a "clover-leaf" pattern) are distributed onto fresh plates of culture medium in the order determined by the random number sequence. The process is repeated for each row of the matrix (i.e., each replicate). This (or any similarly devised) randomization process effectively reduces the number of replicates required to see differences between treatments in a single experiment.

In summary, this experimental design calls for bombardment of either two seedlings or four kernel halves per treatment. With three replicates, an experiment involving 15 treatments would require the dissection of 75 seeds, which is not too formidable after a little practice.

Data Collection and Analysis

Both the onset of anthocyanin accumulation and the rate at which anthocyanins accumulate vary from cell to cell following the introduction of plasmids expressing R and/or C. In some cells, pigment can be detected as soon as 6 h after bombardment, but in others, pigment will only be visible after 48 h. One should therefore wait 2–3 days before enumerating the total number of red cells in most targets; at this point, very few cells will be missed. Cells containing anthocyanins will stay pigmented for a considerable period, so it is possible to score experiments later, if necessary. At the time of counting, I include any cell containing anthocyanin pigment detectable by eye. A dissection microscope with appropriate illumination is essential. An eyepiece graticule and/or a tally counter are also useful. I have recently started using a custom-built counter that can be operated with one's feet; this leaves one's hands free for focusing and moving the field of view. Finally, it is a good idea to estimate one's own accuracy by counting at least one sample several times: the coefficient of variation (CV or standard deviation/mean) should be 5% or less.

With each of the experimental systems described above, variation in both the immediate prehistory and the exposed surface area of the target tissue can be kept to a minimum. However, the CV for red cell numbers in a small number of replicate samples is frequently 30% or higher. This is because some factors (e.g., number of particles that actually enter cells, positioning of the "zone of death," etc.) are more difficult to control (Klein *et al.,* 1988). It is important, therefore, to design experiments in which variation between treatments can be distinguished effectively from random variation. One way to analyze the data is to assume that the variation is log-normally distributed, and then to convert all red cell numbers to their natural logarithm values prior to statistical analysis. However, when using a wide range of plasmid doses, it is often easier to sum replicate cell counts and graph the dose-response data on a log–log plot (see below).

Results and Discussion

Some typical dose-response data for pPHI443, a 35S :: R construct (Ludwig *et al.,* 1990), are shown in Figure 1A. Several features can be noted: (1) at low doses of 35S :: R, the number of red cells seen in a fixed number of bombardments depends on the amount of input template; (2) at high doses of 35S :: R, the number of red cells apparently decreases; and (3) the optimum level of 35S :: R is different in aleurones and seedlings. When the data are graphed on a log–log plot (Figure 1B), the response profiles are fairly linear over a wide range of input DNA levels. These features are consistent with a model in which (1) only a small number of 35S :: R templates are active per cell (Weintraub, 1988; Armaleo *et al.,* 1990) and (2) a threshold number of 35S :: R templates is

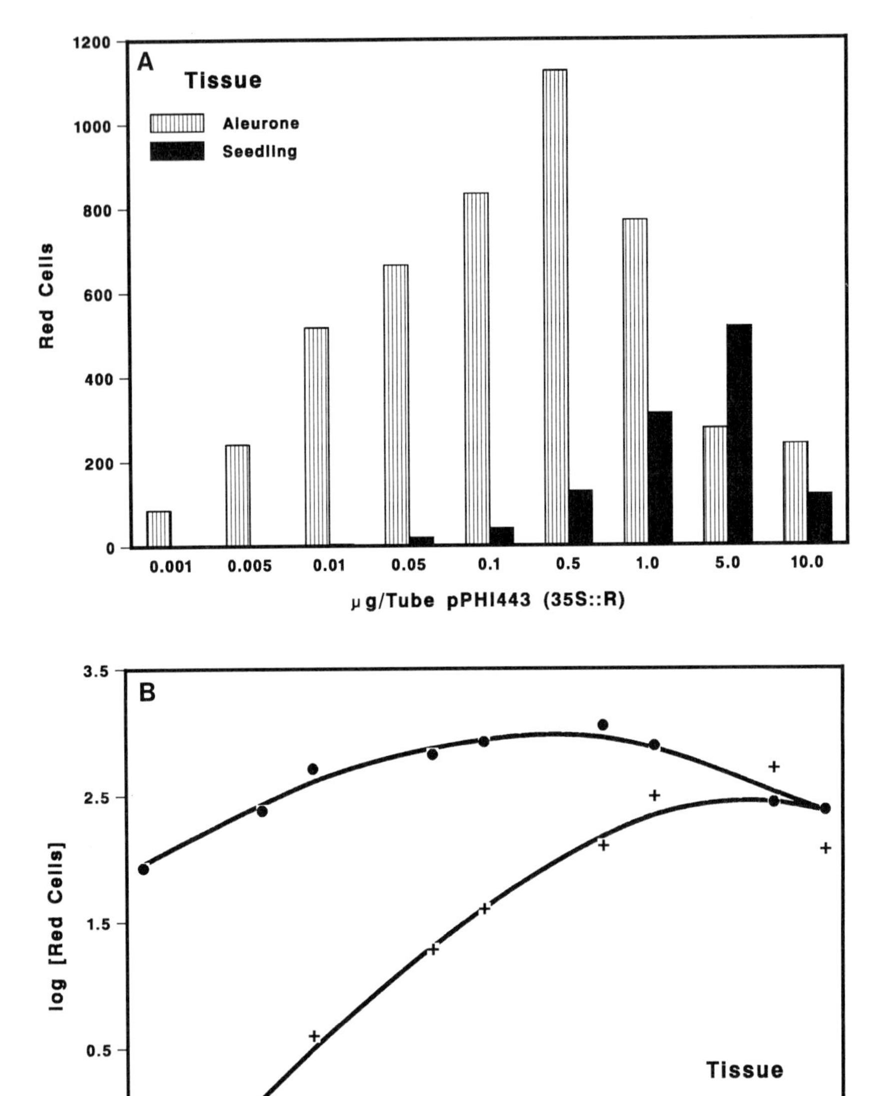

Fig. 1 (A) Dose-response data for pPHI443 (35S :: R) in aleurone cells and seedlings. Red cell numbers shown are the sum of four replicate treatments. Each tube contained the indicated amount (x) of pPHI443, ($10 - x$) μg of pPHI459 (35S :: GUS), and 4.375 mg of tungsten particles. (B) Same data as in (A) graphed as a log–log dose-response plot.

required before the anthocyanin pathway is activated. Increasing the amount of input 35S :: R increases the probability that a cell will contain the required threshold number of templates. Similar inferences can also be made from 35S :: C dose-response data (not shown).

In order to make comparisons between treatments involving R and/or C, it is important to use several doses of input template that fall within a range where the log–log dose-response curve is approximately linear. The slope of the log–log dose-response curve obtained with several doses of template is more informative than the absolute number of red cells at any given dose. This is best illustrated by examining the results of an experiment addressing whether two constructs with different promoter sequences show tissue-specific differences in gene expression (Figure 2). The plasmids pPHI471, pPHI493, and pPHI667 each have a fragment of the 35S promoter from position −90 fused to the R coding sequence (P. Solan, unpublished data). The plasmids pPHI471 and pPHI493 have, in addition, either a duplicated 35S upstream region or the region upstream from the TATA box of a maize phosphoenolpyruvate (PEP) carboxylase gene (Hudspeth and Grula, 1989), respectively. In seedlings, pPHI493 performs better than pPHI471 at all doses tested and the slope of the log–log plot is shallower (Figure 2A). In aleurones, there is little difference between pPHI493 and pPHI471 at the highest dose tested, but the slope of the pPHI493 curve is about fourfold steeper than the pPHI471 curve (Figure 2B). The slope of the dose-response curve for pPHI667 in aleurones is about fivefold steeper than the pPHI471 curve. In green seedlings, therefore, the PEP carboxylase construct, pPHI493, appears to perform marginally better than the enhanced 35S construct, pPHI471, but in aleurones, pPHI493 performs more similarly to the minimal 35S promoter construct, pPHI667. This correlates with the expected tissue-specificity for PEP carboxylase expression in maize (Hudspeth and Grula, 1989).

The effect of the maize *Adh1* intron and the leader sequence from the coat protein mRNA of tobacco mosaic virus on 35S :: R expression in seedlings illustrates these principles further. Both sequence elements have been shown to improve the expression of other reporter genes in maize cells (Callis *et al.*, 1987; Gallie *et al.*, 1989). Two 35S :: R constructs that possess these sequences (pPHI443 and pPHI471) were compared with a third construct (pPHI293) that does not. At the highest dose tested, there was no significant difference between the number of red cells obtained with each of the three plasmids (Figure 3A). At lower doses, however, the response with pPHI293 was significantly reduced. An inappropriate leader sequence can also affect the expression of 35S :: C in a similar fashion. For example, the two 35S :: C constructs,

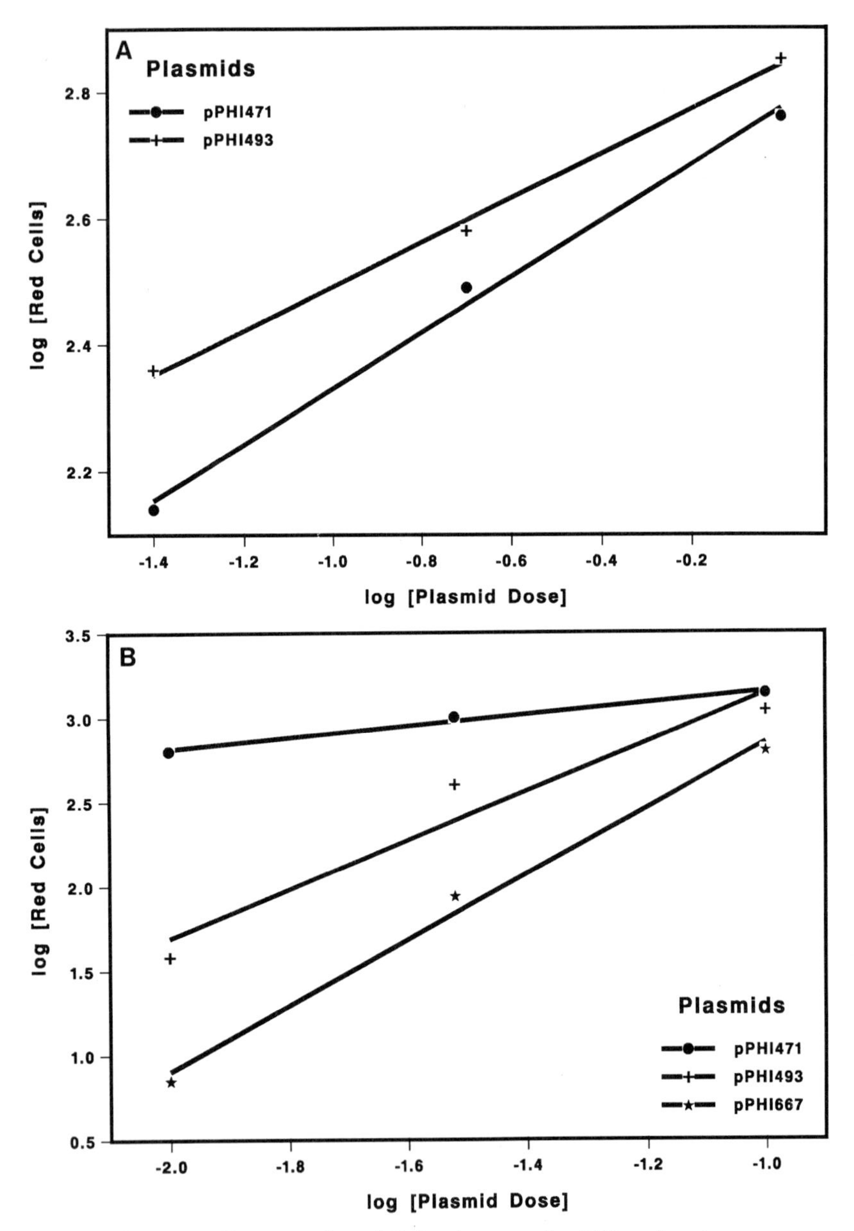

Fig. 2 Tissue specific expression of R fused to a maize PEP carboxylase sequence (Hudspeth and Grula, 1989). Log–log dose-response curves are plotted as in Fig. 1B for pPHI471 and pPHI493 in illuminated green seedlings (A) and for pPHI471, pPHI493, and pPHI667 in aleurone cells (B). Each data point is the sum of three replicate treatments. Plasmids are described in the text.

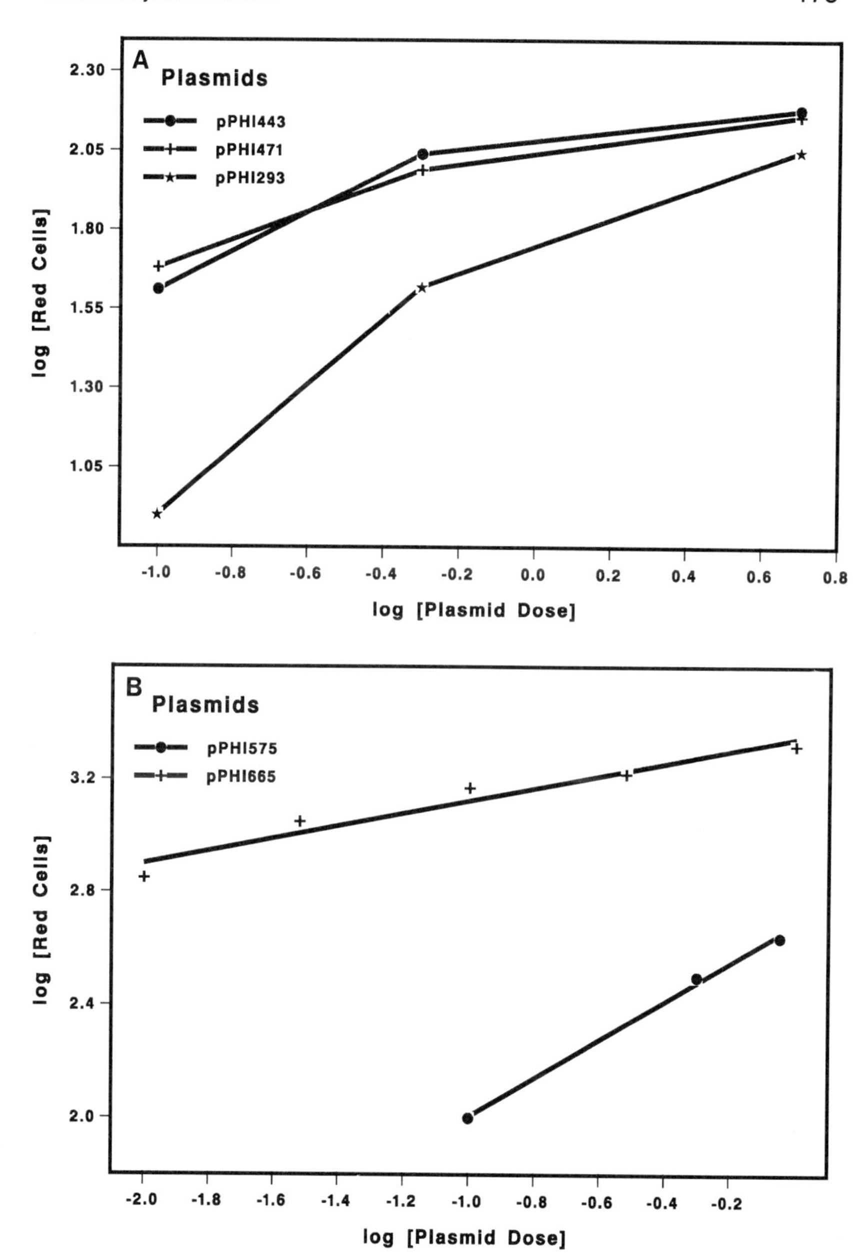

Fig. 3 Effect of different leader sequences on (A) 35S :: R and (B) 35S :: C expression in illuminated green seedlings. Log–log dose-response curves are plotted as in Fig. 1B. Each data point is the sum of three replicate treatments. In (B), all treatments included 35S :: R (0.1 μg/tube pPHI471). Plasmids are described in the text.

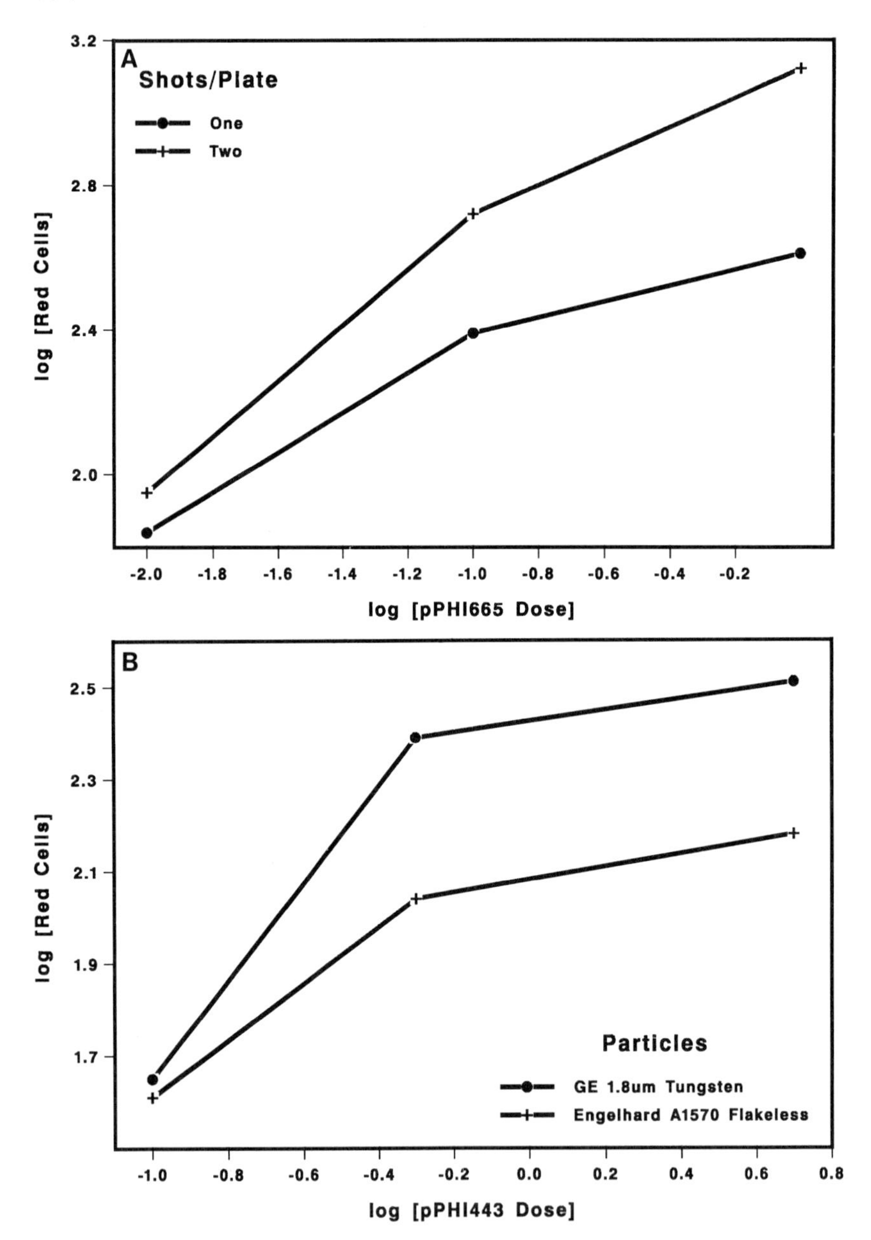

Fig. 4 (A) Effect of one and two shots on pPHI665 (35S :: C) expression in embryogenic suspension culture cells. All treatments included 35S :: R (0.1 μg/tube pPHI471). (B) Effect of different particles on pPHI443 (35S :: R) expression in illuminated green seedlings. In both (A) and (B), each data point is the sum of three replicate treatments. Log–log dose response curves are plotted as in Fig. 2B.

pPHI575 and pPHI665, are identical except for the presence of an out-of-frame initiation codon in the leader sequence of pPHI575 (L. Sims, unpublished data). In seedlings, both the maxima and the slope of the dose-response curves are significantly different (Figure 3B). Presumably, the mRNA encoded by pPHI665 is translated more efficiently than the mRNA encoded by pPHI575.

Red cell numbers can also be used to distinguish between treatments which affect the delivery of DNA. In the two examples shown in Figure 4, two shots were more effective than one in embryogenic suspension cells (Figure 4A) and 1.8-μm tungsten particles from General Electric were better than Engelhard gold particles for introducing DNA into seedlings (Figure 4B).

These examples serve to illustrate how transient expression studies in maize tissues with R and/or C allow the investigator to make reasonable inferences about DNA delivery and comparative gene expression using the particle gun. Anthocyanin markers can also be used in stable transformation experiments (M. Ross, D. Tomes, and B. Bowen, unpublished data). For example, chimeric maize plants containing sectors stably transformed with R and/or C constructs can be obtained by bombarding embryos early in development. Most of these stable transformation events are visible as small red sectors in one or more of the first few leaves of young seedlings (Color Plate 7E). Occasionally, red sectors extend into the tassel. However, no germline transmission has yet been observed, presumably because none of the transformed sectors has been derived from the same cells that give rise to the pollen.

Acknowledgments

I would like to thank Larry Beach, John Grula, Richard Hudspeth, Steve Ludwig, Brad Roth, Lynne Sims, Pete Solan, Sue Wessler, and Udo Wienand for providing plasmids; Marc Albertsen and Sheila Maddock for providing plant materials; Bruce Drummond, Ted Klein, Margit Ross, Dwight Tomes, and Jim Wong for communicating unpublished results; and Gary Huffman, Renee Kosslak, and Jerry Ranch for comments on the manuscript.

References

Armaleo, D., Ye, G.-N., Klein, T. M., Shark, K. B., Sanford, J. C.. and Johnston, S. A. (1990). Biolistic nuclear transformation of *Saccharomyces cerevisiae* and other fungi. *Curr. Genet.* 17:97–103.

Benfey, P., Takatsuji, H., Ren, L., Shah, D. M., and Chua, N. H. (1990). Sequence requirements of the 5-enolpyruvylshikimate-3-phosphate synthase 5′-upstream region for tissue-specific expression in flowers and seedlings. *Plant Cell* 2:849–856.

Brouillard, R. (1988). Flavonoids and flower colour. In "The Flavonoids, Advances in Research Since 1980" (J. B. Harborne, ed.), Chapman and Hall, London, pp. 525–538.

Callis, J., Fromm, M., and Walbot, V. (1987). Introns increase gene expression in cultured maize cells. *Genes Dev.* 1:1183–1200.

Coe, E. H., Neuffer, M. G., and Hoisington, D. A. (1988). The genetics of corn. In "Corn and Corn Improvement," 3d ed., (G. F. Sprague, and J. W. Dudley, eds.) ASA, CSSA, SSSA, Madison, Wis., pp. 81–258.

Gallie, D. R., Lucas, W. J., and Walbot, V. (1989). Visualizing mRNA expression in plant protoplasts: Factors influencing efficient mRNA uptake and translation. *Plant Cell* 1:301–311.

Goff, S. A., Klein, T. M., Roth, B. A., Fromm, M. E., Cone, K. C., Radicella, J. P., and Chandler, V. L. (1990). Transactivation of anthocyanin biosynthetic genes following transfer of B regulatory genes into maize tissues. *EMBO J.* 9:2517–2522.

Hudspeth, R. L., and Grula, J. W. (1989). Structure and expression of the maize gene encoding the phosphoenolpyruvate carboxylase isozyme involved in C4 photosynthesis. *Plant Mol. Biol.* 12:579–589.

Kiesselbach, T. A. (1980). "The Structure and Reproduction of Corn." University of Nebraska Press, Lincoln.

Klein, T. M., Gradziel, T., Fromm, M. E., and Sanford, J. C. (1988). Factors influencing gene delivery into *Zea mays* cells by high-velocity microprojectiles. *Bio/Technology* 6:559–563.

Klein, T. M., Roth, B. A., and Fromm, M. E. (1989). Regulation of anthocyanin biosynthetic genes introduced into intact maize tissues by microprojectiles. *Proc. Natl. Acad. Sci. USA* 86:6681–6685.

Ludwig, S. R., Bowen, B., Beach, L., and Wessler, S. R. (1990). A regulatory gene as a novel visible marker for maize transformation. *Science* 247:449–450.

Ludwig, S. R., and Wessler, S. R. (1990). Maize R gene family: tissue-specific helix-loop-helix proteins. *Cell* **62:**849–851.

Menssen, A., Hohmann, S., Martin, W., Schnable, P. S., Peterson, P. A., Saedler, H., and Gierl, A. (1990). The En/Spm transposable element of *Zea mays* contains splice sites at the termini generating a novel intron from a dSpm element in the A2 gene. *EMBO J.* 9:3051–3057.

Meyer, P., Heidman, I., Forkman, G., and Saedler, H. (1987). A new petunia flower colour generated by transformation of a mutant with a maize gene. *Nature* 330:677–678.

Mol, J. N. M., Stuitje, A. R., Gerats, A. G. M., and Koes, R. E. (1988). Cloned genes of phenylpropanoid metabolism. *Plant Mol. Biol. Rep.* 6:274–279.

Napoli, C., Lemieux, D. and Jorgensen, R. (1990). Introduction of a chimeric chalcone synthase gene into petunia results in reversible co-suppression of homologous genes in *trans*. *Plant Cell* 2:279–289.

Nash, J., Luehrsen, K. R., and Walbot, V. (1990). Bronze-2 gene of maize: Reconstruction of a wild-type allele and analysis of transcription and splicing. *Plant Cell* 2:1039–1049.

Sambrook, J., Fritsch, E. F., and Maniatis, T. (1989). "Molecular Cloning: A Laboratory Manual," 2nd ed. Cold Spring Harbor Laboratory Press, Cold Spring Harbor, N.Y.

van der Krol, A. R., Lenting, P. J., Veenstra, J. G., van der Meer, I. M., Koes, R. E., Gerats, A. G. M., Mol, J. N. M., and Stuitje, A. R. (1988). An "antisense chalcone synthase gene" in transgenic plants inhibits flower pigmentation. *Nature* 333:866–869.

van der Krol, A. R., Mur, L. A., Beld, M., and Stuitje, A. R. (1990). Flavonoid genes in petunia: addition of a limited number of gene copies may lead to a suppression of gene expression. *Plant Cell* 2:291–299.

Weintraub, H. (1988). Formation of stable transcription complexes as assayed by analysis of individual templates. *Proc. Natl. Acad. Sci. USA* 85:5819–5823.

PART

5 Applications of GUS to Animal Genetic Analysis

13 GUS as a Useful Reporter Gene in Animal Cells

Daniel R. Gallie
Department of Biochemistry
University of California
Riverside, California

Virginia Walbot
Department of Biological Sciences
Stanford University
Stanford, California

John N. Feder
Howard Hughes Medical Institute, and
Department of Physiology and Biochemistry
University of California
San Francisco, California

Until recently, virtually all investigations employing the *Escherichia coli uidA* (currently referred to as *gusA*) locus as a reporter for expression studies have been limited to higher plants. The low endogenous β-glucuronidase (GUS) activity present in most angiosperms has allowed the *uidA* gene to be used as a sensitive marker. The *uidA* gene has not gained wide usage among investigators working with animal tissues, primarily because many higher and lower animals contain endogenous GUS activity. In fact, most commercially available preparations of GUS protein originate from animal tissue, for example, from mollusks or bovine liver. As a result, the most popular reporter genes in animal studies are chloramphenicol acetyltransferase (CAT) and firefly luciferase (LUC). The assays for these reporter genes can entail considerable expense, effort, and exposure to radioisotopes.

By far the most widely used reporter gene to date is CAT. Very little endogenous chloramphenicol acetyltransferase activity is found in most animal tissues tested, making it a reasonably sensitive reporter gene. Moreover, the CAT enzyme is stable at 37°C. The assay for CAT activity, particularly for low levels of expression, is rather involved,

GUS Protocols: Using the GUS Gene as a Reporter of Gene Expression

usually requiring 1 day to process the samples and a TLC plate plus a second day to expose the TLC plate to film. In addition, the assay is relatively expensive and potentially hazardous as it requires working with organic solvents in the TLC analysis as well as ^{14}C-labeled chloramphenicol.

More recently, the LUC gene has been used in mammalian cells (de Wet *et al.*, 1987). The assay for luciferase is both easy and fast; as there is virtually no background (in this case, light) in either plant or animal tissues, LUC is a more sensitive reporter gene than either CAT or GUS. Both the substrate, luciferin, and a luminometer, required to detect initial photon production, are expensive. Although the luciferase enzyme has a half life of approximately 15 h at 23°C in plant cells, at the temperature normally used for animal tissue culture, it is considerably shorter (approximately 40 min in Chinese hamster ovary cells). This can be a significant drawback if the stability of the marker protein is important.

The GUS reporter gene is an excellent alternative: the assay requires slightly more effort than the LUC assay; the sensitivity is excellent; the bacterial GUS enzyme is stable at 37°C; the assay is both inexpensive and safe. Moreover, fluorimeters that are exactly suited to detecting the fluorogenic product of the GUS reaction with 4-methylumbelliferyl β-D-glucuronide are now available and are far less expensive than a luminometer.

Use of *uidA* gene as a reporter gene in animal studies has therefore been unduly overlooked. Although these animals do have β-glucuronidase-encoding genes and GUS activity can be detected in many tissue types, these genes are not so highly expressed as to represent a barrier to the use of the *uidA* gene as a marker gene. Indeed, a recent study in mouse (Bracey and Paigen, 1987) indicated that tissues of this species contain approximately two molecules of GUS mRNA per cell. Moreover, many of the endogenous GUS animal enzymes have pH optima in the acidic range, from 3.8 to 5.0, whereas the assay for the GUS encoded by the *uidA* gene is carried out at a pH of 7.4. It should be made clear, however, that the background in Chinese hamster ovary cells (and probably most other tissue sources) is higher than the background from cells from tobacco or maize. This notwithstanding, we have found the *uidA* gene to be an excellent reporter gene for our studies in animal cells.

We have used the *uidA* gene to study posttranscriptional regulation of gene expression in animal cells. We are particularly interested in elucidating the determinants within a given messenger RNA that influence the stability and the translational efficiency of an mRNA. Our approach involves assaying transient expression from mRNA constructs introduced into cells using electroporation. This method is an extremely efficient delivery system that transforms virtually 100% of the cells (Gallie *et al.*, 1989).

To focus specifically on the regulation involving mRNA, that is, regulation occurring solely in the cytoplasm, we avoid using DNA-based constructs as expression from these would require, subsequent to electroporation, partitioning of the construct into the nucleus, followed by transcription, processing of the pre-mRNA, and nucleocytoplasmic transport of the mature mRNA. All of these steps must occur before the mRNA is subject to the types of regulation that specifically interest us. The multiplicity of steps associated with using DNA-based constructs may well contribute to overall error associated with DNA-based transient expression analysis. Moreover, when DNA-based constructs are mutagenized, alterations in the transcribed region of the constructs that result in differences in expression could reflect altered mRNA stability, translational efficiency, promoter activity, processing signals, or mRNA transport. To identify the precise cause of altered expression is therefore a complex task. For example, to determine mRNA half-life, transcription must be inhibited and the decay kinetics of the existing mRNA must be followed. In general, unless the mRNA is very short-lived, the agents used (most often actinomycin D) are so toxic to the cell that prolonged exposure precludes firm conclusions. Moreover, the effect of actinomycin D can be pleotropic and may affect more than just the rate of transcription.

A more direct way to measure posttrancriptional contributions to gene regulation is to introduce mRNA directly into the cytoplasm of a cell. By so doing, not only is the mRNA delivered to the appropriate cellular compartment but any potential complications from transcription or pre-mRNA processing are avoided. As a result, changes in expression from alterations in the construct must result from changes in either message stability or its translational efficiency. With the development of *in vitro* synthesis from either SP6- or T7-based vectors, it is possible to control exactly what sequences are present at the termini of a given mRNA. Therefore, it is possible to study alterations in an mRNA that cannot otherwise be produced *in vivo* but may be important in understanding posttranscriptional regulation, such as the role of the length of the poly(A) tail, the role of other sequences that may functionally replace a poly(A) tail, and the impact of cap (GpppN) methylation.

The *uidA* gene has proven to be particularly useful in our study of the role the 5′- and 3′-untranslated regions (UTR) play in posttranscriptional regulation. For example, we observed in our study of the poly(A) tail that this mRNA determinant is reporter-gene dependent. The addition of a poly(A) tail to GUS mRNA improved expression 5–10 times more than did the addition of the same poly(A) tail to LUC mRNA (compared to constructs lacking a tail). In contrast, the 5′-UTR from tobacco mosaic virus, a sequence that acts as a translational enhancer, is reporter-gene independent, that is, this sequence enhances to an equal extent the translation of either GUS or LUC mRNA. Therefore, for the study of mRNA determinants involved in regulating gene expression that are located 3′ to the open reading frame of a gene, the *uidA* gene is a far more sensitive reporter gene than LUC.

Procedure: In Vitro *Synthesis of RNA*

1. Linearize the T7-based plasmid DNAs with the appropriate restriction enzymes: for T7-GUS, EcoRI is used; for T7-GUS-A_{25} and T7-GUS-A_{50}, DraI is used.
2. Enough DNA for a final concentration of 100 μg/ml is added to the *in vitro* synthesis reaction containing 100 μg/ml bovine serum albumin; 500 μM ATP, CTP, and UTP; 160 μM GTP; 1.6 M m^7GpppG; 40 mM Tris-HCl, pH 7.5; 6 mM $MgCl_2$; 10 mM dithiothreitol; 250 U/ml RNasin; 1000 U/ml T7 RNA polymerase. This mixture is incubated at 37°C for 2 h. The stock solutions need not be made with DEPC-treated water if well-autoclaved water is used. If there is any doubt about the quality of the water, however, then it should be DEPC-treated before autoclaving. (Note: It is important that the ribonucleotide stocks be adjusted to a neutral pH before adding them to the transcription reaction, as they will otherwise acidify the reaction and hence lower the yield of RNA product.)
3. To terminate synthesis add EDTA to a final concentration of 10 mM in the reaction mix. Extract the reaction once with a 1 : 1 phenol : chloroform mixture and once with an equal volume of ether.
4. Precipitate the RNA by the addition of ammonium acetate to a final concentration of 0.7 M followed by an equal volume of isopropanol. Incubate on ice for 1 h.
5. Centrifuge the solution for 15 min at 10,000 × g in a microcentrifuge. Remove the isopropanol mixture and wash the pellet extensively with 80% ethanol in water. The RNA pellet can be stored under 100% ethanol at −80°C indefinitely or resuspended in autoclaved, distilled water and used immediately.

These reaction conditions will result in a high percentage of capped transcripts. A 1:10 ratio of the GTP : m^7GpppG should be maintained so that the T7 RNA polymerase is most likely to initiate transcription with a cap. The level of GTP is usually reduced with respect to the other nucleotides to lower the cost of RNA synthesis; the m^7GpppG is the most expensive reagent. It is vital that the synthesized RNA be capped as this is one of the most essential determinants regulating the efficiency of expression. The template DNA is usually not removed from the newly synthesized RNA as it does not normally serve as a template for expression in plant or animal cells: the T7 promoter does not function in eukaryotes and there is no eukaryotic promoter in the constructs we employ. An amount of DNA equivalent to that used in the transcription reactions can be electroporated into cells to serve as a control to demonstrate that no expression results from the DNA alone. We believe this is a safer approach than removing the DNA using DNase. DNase preparations are sometimes slightly contaminated with RNases, and the removal of even just a few bases from the 5′ termini of mRNAs results in the loss of the cap and has a profound effect on expression.

The RNA produced can be quantitated by several means. If several mRNA constructs are to be compared and a relative quantitation is sufficient, a small portion of each RNA can be displayed on a denaturing formaldehyde-agarose gel and stained with ethidium bromide. Visual inspection can usually detect differences between samples that are greater than 20%. A second method is to trace label the synthesized RNA by adding 5–10 μCi UTP or CTP to the transcription reaction. The RNAs can then be quantitated by counting a small portion of each purified RNA in a scintillation counter. The RNA may also be quantitated spectrophotometrically. If DNA is present in the RNA, then an equivalent amount of DNA to that present in the RNA sample should be measured separately and then subtracted from the RNA + DNA sample to obtain the concentration of the RNA.

As the electroporation of Chinese hamster ovary (CHO) cells with RNA had not been attempted before we initiated these studies, it was necessary to determine appropriate paramenters for efficient delivery of mRNA. To accomplish this, we employed LUC mRNA as we had not discovered that the *uidA* gene could be employed in animal cells. In retrospect, however, these experiments could have been carried out just as easily using GUS mRNA as the test RNA.

The mammalian cell line used, CHO K1 (Kao and Puck, 1967), was maintained in Ham's F-12 medium (Gibco) supplemented with 10% fetal bovine serum (Gibco). Cells were cultured for electroporation by inoculating a 175-cm^3 flask (Falcon) with approximately 1×10^6 cells, and these were grown to a final density of 5 × 10^6/flask (3 × 10^4/cm^2). The cells were harvested by trypsinization [0.05% trypsin, 0.53 m*M* ethylenediamine tetraacetic acid (EDTA)], washed once with 10 ml of complete medium, then resuspended in phosphate-buffered saline (PBS; Gibco) to a concentration of 6 × 10^6/ml. A 0.8-ml aliquot (approximately 3–5 × 10^6 cells) was mixed with the test RNA that had been resuspended in 50 μl water; the CHO cells were immediately placed in a electroporation chamber followed by discharge of the capacitor (Gene Pulser; Bio-Rad). Following electroporation, the cells were kept at room temperature for 10 min and then 10 ml of Ham's medium was added to the cells. The cells were incubated at 37°C in 5% CO_2 for 16 h after which the media was aspirated off, and the plates were gently washed twice with cold PBS. The cells were removed from the plate by gentle scraping using a rubber policeman in 5 ml of PBS. They were then harvested by centrifugation prior to resuspension in GUS assay buffer and sonicated for 5 s. It is important that the cells be thoroughly washed before the addition of the test RNA, because serum may contain RNases. Well-washed cells are virtually free of these RNases. Although we normally electroporate the cells immediately following the addition of the test RNA, we have observed little difference in the resulting expression if electroporation occurs 1 min after addition of the test RNA.

With the BioRad Gene Pulser, we electroporated 1 μg mRNA, using a range of voltages and capacitances, and followed the resulting level of LUC expression as a measure of the efficiency of delivery. When the capacitance was constant and RNA was electroporated using voltages ranging from 625 to 1125 V/cm, the resulting expression continued to increase with increased voltage. Similar results were obtained when the voltage was held constant and the capacitance increased from 25 to 960 μF. Although expression was highest under conditions of high voltage or capacitance, cell recovery was progressively inhibited. Any data resulting from cells in a nonphysiological state cannot be viewed as reflecting the true status of the cell. In order to obtain transformed cells that were both healthy and had received test mRNA, the electroporation conditions were set at 350–450 V with 250 μF capacitance.

The dose response of RNA electroporation with CHO cells was also examined, as it is important to deliver an amount of RNA to the cells that does not overwhelm the translation machinery and thereby potentially obscure the normal competition for initiation factors, ribosomes, etc. Expression increased as a linear function of input mRNA up to at least 20 μg mRNA. Amounts greater than 20 μg mRNA were not examined, as this quantity far exceeds the amount required to obtain a high level of expression. For typical electroporations with GUS mRNA, 0.5–2.0 μg of mRNA was used. The error associated with mRNA electroporation as measured from several replicate electroporations with the same cell preparation was ±15%.

Figure 1 illustrates the GUS mRNA constructs tested for expression in CHO cells. The GUS mRNA construct had been produced from DNA linearized with EcoRI, whereas the GUS-A_{25} and GUS-A_{50} mRNAs had been synthesized from DNA constructs linearized with DraI. Approximately 1 μg mRNA of each construct was used for the electroporation.

When GUS mRNA was introduced into CHO cells, the resulting level of expression (1.4 U) was approximately five times above background (0.3 U). In contrast, the same GUS construct electroporated into tobacco protoplasts yielded a level of expression that was not above background (0.01 U). From an examination of the background activities of the plant and animal cells tested, there is significantly more (30-fold) endogenous GUS activity in the CHO cells than in tobacco. Offsetting this greater background, however, is the much higher expression of GUS from the input mRNA (140-fold) in CHO compared to the tobacco cells.

The addition of a poly$(A)_{25}$ tail to the GUS mRNA increased expression 14-fold in CHO cells and 34-fold in tobacco protoplasts. A greater increase in expression was observed in both cell types when a poly$(A)_{50}$ tail was added. The level of GUS expression in CHO cells for polyadenylated mRNA, therefore, was up to 100 times above background, and for tobacco up to 70 times above background. Thus, even though the endogenous activity is higher in the animal cells, the overall level of

Construct	Structure	GUS Specific Activity (nmoles/min/mg) CHO	Tobacco
GUS	GpppG —17 b— AUG — GUS — UGA —67 b— GGGUACCGAGCUCGAAUU-3'	1.4	0.01
GUS-$(A)_{25}$ DraI	GpppG —17 b— AUG — GUS — UGA —67 b— AU$(A)_{25}$UUU-3'	20.0	0.34
GUS-$(A)_{50}$ DraI	GpppG —17 b— AUG — GUS — UGA —67 b— AU$(A)_{25}$GUU$(A)_{25}$UUU-3'	30.0	0.71

Fig. 1 GUS constructs used for electroporation of CHO and tobacco cells. Approximately 1 μg of each construct was used for electroporation. See text for details.

expression is proportionately higher, and as a result, the differential between the background and the expression resulting from the test RNA is as great in the CHO cells as in tobacco. This differential can be easily increased in either cell type by simply increasing the input RNA. Because the relationship between input RNA and expression is directly proportional, doubling the input RNA yields roughly a doubling in the expression. If a 1000-fold differential between background and the level of expression resulting from the GUS-A_{50} was desired, this could be achieved by using 10 μg mRNA instead of the 1 μg employed here.

From these experiments using GUS in CHO cells, we conclude that GUS is a useful reporter gene. Moreover, its utility should not be limited to mRNA electroporation. Because expression from DNA-based constructs tends to be greater than mRNA-based constructs when compared on a per microgram basis, GUS should prove to be an equally good reporter gene for those studies of the transcriptional control of gene expression. The use of *uidA* in other mammalian cell lines will depend on the differential between the endogenous GUS activity and the level of expression achieved from the particular DNA- or RNA-based construct. If the differential is as great as that found for CHO cells, then the GUS reporter gene will be an excellent alternative to the existing reporter genes.

References

Bracey, L. T., and Paigen, K. (1987). Changes in translational yield regulate tissue-specific expression of β-glucuronidase. *Proc. Natl. Acad. Sci. USA* 84, 9020–9024.

de Wet, J. R., Wood, K. V., DeLuca, M., Helinski, D. R., and Subramani, S. (1987). Firefly luciferase gene: Structure and expression in mammalian cells. *Mol. Cell Biol.* 7, 725–737.

Gallie, D. R., Lucas, W. J., and Walbot, V. (1989). Visualizing mRNA expression in plant protoplasts: Factors influencing efficient mRNA uptake and translation. *Plant Cell* 1, 301–311.

Kao, F., and Puck, T. T. (1967). Genetics of somatic mammalian cells. IV. Properties of Chinese hamster cell mutants with respect to the requirements for proline. *Genetics* 55, 513–524.

14 β-Glucuronidase (GUS) Assay in Animal Tissue

John W. Kyle
Department of Medicine
Section of Cardiology
University of Chicago
Chicago, Illinois

Jeffrey H. Grubb
Department of Biochemistry
and Molecular Biology
St. Louis University
School of Medicine
St. Louis, Missouri

Nancy Galvin and Carole Vogler
Department of Pathology
St. Louis University
School of Medicine
St. Louis, Missouri

Mammalian β-glucuronidase is one of the most extensively studied lysosomal acid hydrolases. It has been studied as a model of enzyme transport and sorting to lysosomes via the mannose 6-phosphate-dependent pathway, as an example of a hormonally and developmentally regulated enzyme, and as an important component in the stepwise degradation of glycosaminoglycans. In addition, the human genetic deficiency of β-glucuronidase results in the lysosomal storage disease mucopolysaccharidosis type VII (MPS VII or Sly syndrome). The recent discovery of β-glucuronidase-deficient mice, which closely resemble humans with MPS VII, presents an excellent animal model with which to test different therapeutic approaches. In this chapter we will briefly review mammalian β-glucuronidase and present four protocols that we use to quantitate and localize β-glucuronidase activity. The first protocol assays for total β-glucuronidase activity fluorometrically by using 4-methylumbelliferyl β-D-glucuronide. The second pro-

GUS Protocols: Using the GUS Gene as a Reporter of Gene Expression

tocol describes how to differentiate isoforms of β-glucuronidase by use of nondenaturing polyacrylamide gel electrophoresis (PAGE) activity gels stained for activity with naphthol AS-BI β-D-glucuronide. Finally, the third and fourth protocols outline two histochemical techniques for the cellular localization of β-glucuronidase in tissues by activity staining with either naphthol AS-BI β-D-glucuronide or 5-bromo-4-chloro-3-indoyl β-D-glucuronide.

Structure and Properties

β-Glucuronidase (β-D-glucuronoside glucuronosohydrolase, EC 3.2.1.31) is found in virtually all mammalian tissues (for review, see Paigen, 1989). The highest known source is the rat preputial gland, where it has been estimated to constitute as much as 5% of the total protein (Himeno *et al.*, 1975). Other tissues with high activity include kidney, liver, and spleen. β-Glucuronidase is a tetrameric glycoprotein composed of four identical subunits. The monomers are synthesized as precursors of approximately M_r = 75,000–82,000 and are proteolytically processed at the carboxy terminus either prior to or soon after reaching the lysosome. β-Glucuronidase is predominantly localized within the lysosome. However, in the liver and kidney of mice and several other mammalian species, β-glucuronidase has a dual localization within lysosomes and the endoplasmic reticulum, where it is bound to the resident ER protein, egasyn (Fishman *et al.*, 1967a; Swank and Paigen, 1973).

β-Glucuronidase acts in the acidic environment of the lysosome as an exoglycosidase to remove terminal β-glucuronic acid residues from the nonreducing end of glycosaminoglycans and other glycoconjugates. Natural substrates include chondroitin sulfate, dermatan sulfate, hyaluronic acid, and heparan sulfate (for review, see Neufeld and Meunzer, 1989). Other potential substrates are β-glucuronic acid-containing glycoproteins, glycolipids, and glucuronic acid conjugated to steroids and certain drugs. Synthetic substrates used to assay β-glucuronidase colorimetrically include phenophthalein β-D-glucuronide and p-nitrophenyl β-D-glucuronide (Fishman *et al.*, 1967b). The most common synthetic substrate is the fluorogenic 4-methylumbelliferyl β-D-glucuronide (see Protocol 1, this chapter; e.g., Glaser and Sly, 1973). Naphthol AS-BI β-D-glucuronide has been used for histochemical staining of tissue sections (Hayashi *et al.*, 1968; see also Protocol 3, this chapter) and for staining nondenaturing PAGE gels (e.g., Glaser *et al.*, 1975; Kyle *et al.*, 1990; see also Protocol 2, this chapter).

Mammalian β-glucuronidase has a broad pH optimum ranging between 3.8 and 5 (e.g., Brot *et al.*, 1978). The acidic pH optima for mammalian β-glucuronidase permits it to be distinguished from *Escherichia coli* β-glucuronidase, which has a neutral pH optimum. One interesting feature of β-glucuronidase is resistance to heat denaturation. For example, human β-glucuronidase is almost completely stable to heating at 65°C for 90 min. Differential sensitivity to thermal denaturation has been useful in assays where it is necessary to discriminate between β-glucuronidases from different species (e.g., Oshima *et al.*, 1987).

Molecular Biology

β-Glucuronidase cDNAs have been isolated from several sources including human (Oshima *et al.*, 1987), rat (Nishimura *et al.*, 1986; Powell *et al.*, 1988), and mouse (Funkenstein *et al.*, 1988; Gallagher *et al.*, 1989). The deduced amino acid sequences show a high degree of similarity. The rat amino acid sequence is 77% similar to the amino acid sequence deduced for the human placental clone after excluding the signal sequence (Powell *et al.*, 1988; Miller *et al.*, 1990). The rat and human amino acid sequences are 47% similar to *E. coli* β-glucuronidase (Jefferson *et al.*, 1986; Powell *et al.*, 1988). Rat and human amino acid sequences have weaker homology to regions of *E. coli* β-galactosidase (Nishimura *et al.*, 1986; Oshima *et al.*, 1987; Powell *et al.*, 1988).

Surprisingly, two closely related cDNAs were isolated from human placenta (Oshima *et al.*, 1987). The two cDNAs differed only by a 153-bp deletion within the middle of the coding sequence. Both cDNAs expressed proteins following transfection into Cos-7 cells, although only the larger cDNA produced enzymatically active β-glucuronidase (Oshima *et al.*, 1987). Determination of the genomic organization for the human gene revealed that the 153 bp represents an exon (exon 6) and probably arises from alternate splicing (Miller *et al.*, 1990). Only one type of cDNA has been reported for rat and mouse.

The mouse (D'Amore *et al.*, 1988) and human (Miller *et al.*, 1990) β-glucuronidase genes have been cloned. Both genes contain 12 exons and 11 introns. The murine gene is found on the distal end of chromosome 5 and the human gene is found on chromosome 7.

Targeting of β-Glucuronidase to Lysosomes

Studies of human β-glucuronidase have provided important insights into the mechanism of targeting lysosomal enzymes to lysosomes via the mannose 6-phosphate-dependent pathway (for review, see von Figura and Hasilik, 1986; Kornfeld and Mellman, 1989). In this pathway, mannose 6-phosphate is added to specific oligosaccharide side chains on newly synthesized lysosomal enzymes in a pre-Golgi or early Golgi compartment. The mannose 6-phosphate recognition marker is bound by mannose 6-phosphate (Man 6-P) receptors in the trans Golgi, and the receptor–enzyme complex is translocated to acidic prelysosomal vesicles where the acidic pH causes dissociation of the enzyme from its Man 6-P receptor. The dissociated lysosomal enzymes are transferred to lysosomes and the Man 6-P receptors recycle to the Golgi or cell surface. Mannose 6-phosphate receptors on the cell surface are capable of mediating the endocytosis of extracellular lysosomal enzymes. Endocytosed enzyme is transported to lysosomes through acidic endosomal or prelysosomal compartments.

Mucopolysaccharidosis Type VII (β-Glucuronidase Deficiency)

Severe genetic deficiency of β-glucuronidase leads to a progressive accumulation of undegraded glycosaminoglycans within lysosomes (for review, see Neufeld and Meunzer, 1989). This inherited disorder was originally identified in humans and named mucopolysaccharidosis type VII (MPS VII or Sly syndrome) (Sly *et al.*, 1973). A β-glucuronidase-deficient dog has also been reported (Haskins *et al.*, 1984). Recently, a mouse model for MPS VII was discovered and characterized (Birkenmeier *et al.*, 1989; Vogler *et al.*, 1990). These mice have less than 1% of normal β-glucuronidase levels and have morphological and biochemical characteristics that closely resemble humans with this disorder. Working with Bill Sly in St. Louis and in collaboration with Ed Birkenmeier at Jackson Laboratory in Bar Harbor, Maine, we have begun to access different therapeutic approaches to treatment of MPS VII using the MPS VII mice. These approaches include (1) construction of transgenic mice using the human β-glucuronidase gene (Kyle *et al.*,

1990); (2) bone marrow transplantation (Birkenmeier *et al.*, 1991); and (3) enzyme replacement by infusion of purified recombinant enzyme (Kyle *et al.*, in preparation). These studies require the determination of the total β-glucuronidase activity (see Protocol 1), the distinction between mouse and human β-glucuronidase (see Protocol 2), and the determination of the cellular localization of infused or transplanted β-glucuronidase within a given tissue (see Protocols 3 and 4).

Protocols

Protocol 1: 4-Methylumbelliferyl β-D-Glucuronide (4-MUG) Assay of Mammalian β-Glucuronidase

This protocol can be used as a sensitive fluorometric assay of any mammalian β-glucuronidase. It is adapted from a protocol originally used for screening of MPS VII patients (Glaser and Sly, 1973). Tissue or cells need to be lysed by treatment with either detergent (e.g., 0.5% sodium deoxycholate or 0.5% Saponin) or brief sonication before β-glucuronidase can be assayed.

Reagents

4-MUG assay/substrate buffer:
- 10 m*M* 4-Methylumbelliferyl β-D-glucuronide (4-MUG)
- 0.1 *M* Acetate buffer, pH 4.8
- 1 mg/ml Crystalline bovine serum albumin

Glycine-carbonate stop buffer:
- 0.32 *M* Glycine
- 0.2 *M* Sodium carbonate
- Adjust pH to 10.5

Fluorometer

We use a Farrand fluorometer. The primary filter is a Corning 7-60 with a peak transmittance at 370 nm for excitation. Two emission barrier filters are used: a Kodak Wrattan 2A (transmits light above 415 nm) and a Kodak Wrattan 48 with a peak at 460 nm.

4-MUG Assays

1. Assays are performed in duplicate in 10 × 75 mm borosilicate glass culture tubes. The assay tubes are kept on ice until the start of the incubation. To each tube is added 5–25 μl of sample lysate. Next, 100 μl of 4-MUG assay/substrate buffer is added. The final volume is adjusted to 125 μl and the assays are incubated at 37°C for 30 min to 2 h or more. Multiple time points and sample lysate volumes can be used to ensure that the assay is linear with respect to time and β-glucuronidase concentration, respectively. A substrate blank containing 100 μl of 4-MUG assay/substrate buffer without sample is run with each group of assays as a control. The fluorometer reading obtained from the blank will be subtracted from sample readings. Assay of undiluted cell culture media or other highly buffered samples may require reducing the sample volume or increasing the concentration of acetate buffer (pH 4.8).
2. After incubation at 37°C for the appropriate length of time, the assays are transfered to ice and terminated by the addition of 1.9 ml glycine-carbonate buffer.
3. The fluorometer is calibrated so that 10 nmol of 4-methylumbelliferone (4-MU) in 2 ml of glycine-carbonate buffer gives a reading of 1.00. A standard curve of 4-MU should be read to determine the linear range for the fluorometer. A standard solution of 1 μg/ml quinine sulfate in 0.1 *N* H_2SO_4 can be used to check the fluorometer calibration.
4. The 10 × 75 mm assay tubes are read directly in the fluorometer.

Calculation

One unit of β-glucuronidase activity is defined as the release of 1 nmol 4-MU per hour. β-Glucuronidase activity is calculated as follows:

$$\text{Activity} = \text{units/ml} = \frac{(\text{reading} - \text{blank}) \times 10}{(\text{ml of sample assayed}) \times \text{time (h)}}$$

The factor 10 in the above calculation originates from adjustment of the fluorometer scale to 1 = 10 nmol 4-MU.

Protein concentration is determined by the method of Lowry *et al.* (1951) and specific activity is calculated by dividing units/ml by mg/ml protein.

Protocol 2: Naphthol AS-BI β-D-Glucuronide Staining of Nondenaturing PAGE Gels

We have used the following protocol to distinguish mouse and human β-glucuronidase in transgenic mice (Kyle *et al.*, 1990). We have also used it to examine the formation of hybrid mixed tetramers in Cos-7 cells cotransfected with human and rat β-glucuronidase cDNAs (Fig. 1). A similar protocol has been used by Swank and others to study the association of egasyn with β-glucuronidase (e.g., Medda and Swank, 1985).

Reagents

AS-BI solution : naphthol AS-BI β-D-glucuronide (Sigma): 12.5 mg/ml in dimethylformamide

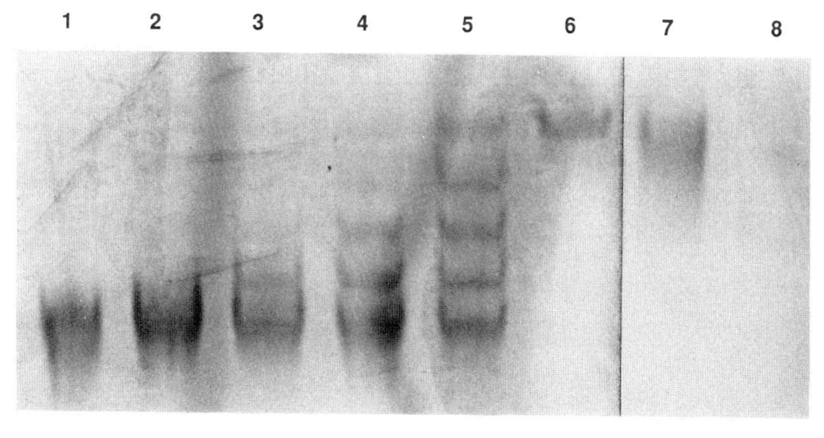

Fig. 1 β-Glucuronidase activity staining in nondenaturing polyacrylamide gels demonstrating the expression of human/rat heterotetramers in transfected Cos-7 cells. Cos-7 cells were transfected with plasmids containing human and/or rat β-glucuronidase cDNA. Transfected cells were scraped from tissue culture dishes (72 h posttransfection) with a plastic scraper, lysed by sonication, and the soluble supernatants were treated with endoglycosidase F to remove carbohydrate side chains before electrophoresis in a 5% nondenaturing PAGE gel. Mixed tetramers could be identified in cells transfected with both human and rat β-glucuronidase cDNAs. Lanes: 1, purified rat preputial β-glucuronidase; 2, Cos-7 cells transfected with rat cDNA only; 3–5, Cos-7 cotransfected with increasing ratios of human/rat cDNA; 6, Cos-7 cells transfected with human cDNA only; 7, human spleen β-glucuronidase; 8, mock transfected Cos-7 cells.

Staining solution

1ml AS-BI solution (12.5 mg/ml)
49 ml 0.2 *M* Acetate buffer, pH 4.8
23 mg Fast garnet GBC salt (Sigma Chemical Co.)
Stir and filter through a 0.45-μm filter.

Sample Preparation

Tissue and cell lysates are prepared without detergent, usually by sonication. For human β-glucuronidase, the heterogeneity of the carbohydrate side chains causes the bands to smear and therefore it is necessary to deglycosylate the samples by treatment with endoglycosidase F. Endoglycosidase F is used without detergent and without denaturing the sample. No enzyme activity is lost by endoglycosidase F treatment.

Deglycosylation

25 μl 0.1 *M* Citrate-phosphate buffer, pH 5.5
1 μl 0.5 *M* EDTA, pH 7.5
5 μl 10 × Protease inhibitors (0.2 mg/ml leupeptin and 1 TIU/ml aprotinin)
1–14 μl Sample lysate (100–2000 U β-glucuronidase)
0.2 U Endoglycosidase F (Boehringer Mannheim)
Add distilled water to a final volume of 50 μl.

Samples are incubated at 37°C for 12–24 h. β-Glucuronidase activity should be assayed before and after treatment to ensure no loss of activity.

Nondenaturing PAGE (Adapted from Davis, 1964)

Electrode buffer (per liter in distilled water)

6 g Tris
28.8 g Glycine
5 % Separating gel (pH 8.9):
3.0 ml 3 *M* Tris-HCl, pH 8.9
6.0 ml 30% Acrylamide-0.8 % bisacrylamide
20.9 ml Distilled water

Degas, then add 100 μl of freshly prepared ammonium persulfate (10%), 10 μl TEMED; mix and pour gel. Add a thin layer of water on top of gel and allow to polymerize (approximately 1 h). Remove water layer before pouring stacking gel.

3 % Stacking gel (pH 6.7)
2.0 ml 0.5 *M* Tris-HCl, pH 6.7
4.0 ml 10% Acrylamide–2.5 % bisacrylamide
8.0 ml 40% Sucrose

Degas, add 2 ml of riboflavin (0.04 mg/ml), pour stacking gel, insert combs, and allow to polymerize under fluorescent light (approximately 15–30 min).

Electrophoresis

Electrophoresis is run at 4°C. The gels are prerun at 25 mA constant current for 15 min. We prerun the gels to remove inhibitors of β-glucuronidase activity, although prerunning will disturb the pH boundary between the stacking and separating gels. Between 100 and 2000 U (nmol/h with 4-MUG) of β-glucuronidase should be used per lane. The deglycosylated sample is mixed 1 : 1 with 40% sucrose and 2 μl of 0.4 % bromophenol blue added as tracking dye. Up to 150 μl of sample can be loaded using 1-cm-wide combs.

Samples are loaded and run into the gel at 25 mA constant current for 30 min and the current is increased to 50 mA. The electrophoresis is stopped when the bromophenol blue tracking dye reaches the bottom of the gel. If better separation of bands is required, the bromophenol blue tracking dye can be run off the gel and the electrophoresis continued for an additional 15–60 min.

After electrophoresis, the gel is removed and soaked in 0.2 *M* acetate buffer (pH 4.8) for 30 min. The acetate buffer is removed and AS-BI staining solution is added. The gel can be incubated at room temperature or at 37°C. The gel should be placed on an orbital shaker set at low speed. Magenta colored bands will appear where β-glucuronidase activity is located. The gel can be incubated until the desired intensity of staining is obtained (usually 1–5 h). The reaction is stopped by removal of the AS-BI staining solution and washing with 7% acetic acid. The gel can be dried on a gel dryer.

Protocol 3: Naphthol AS-BI β-D-Glucuronide Staining in Frozen Sections

Naphthol AS-BI β-D-glucuronide with pararosaniline has been used to stain for β-glucuronidase activity in partially fixed or frozen tissue sections or cells in culture (Fishman *et al.*, 1967a; Lagunoff *et al.*, 1973). Color Plate 8 shows a section of spleen from a β-glucuronidase deficient MPS VII mouse that received an intraperitoneal injection of purified recombinant human β-glucuronidase. The red staining, indicating β-glucuronidase activity, is found on the splenic capsule facing the peritoneum. We use this procedure without fixation in an attempt to retain as much β-glucuronidase activity as possible. Wolfe and colleagues have recently demonstrated the correction of human MPS VII fibroblasts and MPS VII myoblasts by retroviral vector-mediated gene transfer (Wolfe *et al.*, 1989; Smith *et al.*, 1990). Corrected cells were identified by a slightly different procedure by sequential staining with naphthol AS-BI β-D-glucuronide and pararosaniline (Fishman *et al.*, 1967a; Lagunoff *et al.*, 1973).

Tissue Preparation

Tissue is rapidly dissected from animals and frozen at −170°C in isopentane cooled in liquid nitrogen and mounted in Tissue-tek O.T.C embedding medium (Miles, Inc., Elkhart , Ind.). Tissue can be stored in this manner at −80°C until needed. Frozen sections are cut at 7–10 μm thickness and kept at −20°C until ready to stain for activity.

Reagents (Adapted from Bancroft, 1982)

Substrate solution: 10 mg/ml naphthol AS-BI β-D-glucuronide dissolved in dimethyl formamide.

Pararosaniline solution: 2 g pararosaniline hydrochloride in 50 ml of 2 *N* HCl

Veronal acetate buffer (per 500 ml)

- 9.7 g Sodium acetate ($\cdot 3H_2O$)
- 14.7 g Sodium barbiturate
- 500 ml Distilled water

Solution A:
- 0.8 ml Pararosaniline HCl
- 0.8 ml 4% Sodium nitrite
- Mix the pararosaniline and sodium nitrite at room temperature for approximately 15 min.

Solution B:
- 1.0 ml Substrate solution
- 5.0 ml Veronal acetate buffer
- 13.0 ml Distilled water

AS-BI Staining solution: Mix solution A + solution B and filter through Whatman 1 filter paper. Adjust the pH to 4.8 with sodium hydroxide, if necessary.

Methyl green: 2 g Methyl green powder in 100 ml distilled water.

Naphthol AS-BI β-D-Glucuronide Staining

1. Incubate the slides at 37°C in the AS-BI staining solution for 3 h. The incubation time can be varied, depending on the level of β-glucuronidase activity. β-Glucuronidase activity is seen as red.
2. Rinse slides in distilled water for 10 s.
3. Counterstain the slides with methyl green for 5 min.
4. Dehydrate slides in 95% ethanol, absolute ethanol, and clear in xylene. Mount slides with Permount.

Protocol 4: X-Gluc Staining in Fixed Tissue Sections

The following protocol is used to examine the cellular distribution of β-glucuronidase in semithin plastic sections. The procedure is a modification of the one reported by Sanes and co-workers (Sanes *et al.,* 1986) to localize *E. coli* β-galactosidase using X-gal (Lojda, 1970). In Color Plate 9 can be seen a section from transgenic mouse kidney stained with X-gluc. Intense blue staining can be seen in many cells, particularly the renal proximal tubules.

Reagents

X-Gluc: 40 mg/ml in dimethylsulfoxide

X-gluc (5-bromo-4-chloro-3-indoyl-β-D-glucuronide) was obtained from Research Organics (catalog number 1177B)

Staining solution (per milliliter of solution)
20 μl X-gluc (40 mg/ml)
2 μl $MgCl_2$ (1 *M* stock)
40 μl Potassium ferricyanide (200 m*M* stock)
40 μl Potassium ferrocyanide (200 m*M* stock)
820 μl Acetate buffer, pH 4.8
Prepare 5–20 ml depending on amount of tissue.
Acetate buffer: 0.1 *M* acetate buffer, pH 4.8, 150 m*M* NaCl
PBS: 58 m*M* Na_2HPO_4, 18 m*M* KH_2PO_4, 75 m*M* NaCl
Fixatives: 4% paraformaldehyde in phosphate buffer (58 m*M* Na_2HPO_4, 18 m*M* KH_2PO_4, 100 m*M* NaCl, 2 m*M* $CaCl_2$)
Nuclear fast red
0.1 g Nuclear fast red powder
100 ml 5% Aqueous aluminum sulfate
1 Thymol crystal

Fixation

By transcardial perfusion with 4% paraformaldehyde followed by immersion of extracted tissue in the same fixative. Immersion fixation may also be used.

Staining Procedure

1. Rinse fixed tissue once in ice-cold PBS.
2. Cut 50–200 μm sections using a Vibratome.
3. Rinse sections twice in ice-cold PBS.
4. Rinse twice in ice-cold acetate buffer.
5. Add 250–500 μl X-gluc staining solution and incubate covered at room temperature for 1–12 h.
6. Rinse several times in cold PBS to remove staining solution and yellow background.
7. Hard fix sections in 2% glutaraldehyde for 4–20 h.
8. Rinse twice in cold PBS.
9. Dehydrate in 50, 75, and 95% ethanol, followed by infiltration and embedding in glycol methacrylate (Burns and Bretschneider, 1981).
10. Cut semithin sections (1–2 μm thick) of the embedded samples and counterstain with nuclear fast red.

Acknowledgments

We gratefully acknowledge the encouragement and support of William S. Sly, St. Louis. We also thank Benjamin Wu, John Wolfe, and Edward Birkenmeier for helpful discussions and advice. This work was supported by National Institutes of Health grants GM34182 and DK41082.

References

Bancroft, J. D. (1982). Enzyme histochemistry. In "Theory and practice of histological techniques" (J. D. Bancroft and A. Stevens, eds). Churchill-Livingston, Edinburgh.

Birkenmeier, E. H., Davisson, M. T., Beamer, W. G., Ganschow, R. E., Vogler, C. A., Gwynn, B., Lyford, K. A., Maltais, L. M., and Wawrzyniak, C. J. (1989). Murine mucopolysaccharidosis type VII: Characterization of a mouse with β-glucuronidase deficiency. *J. Clin. Invest.* 83, 1258–1266.

Birkenmeier, E. H., Barker, J. E., Vogler, C. A., Kyle, J. W., Sly, W. S., Gwynn, B., Levy, E., and Pegors, C. (1991). Increased life span and correction of metabolic defects in murine MPS VII following syngeneic bone marrow transplantation. *Blood,* in press.

Brot, F. E., Bell, C. E., and Sly, W. S. (1978). Purification and properties of β-glucuronidase from human placenta. *Biochemistry* 17, 385–391.

Burns, W. A., and Bretschneider, A. (1981). "Thin Is In: Plastic Embedding of Tissues for Light Microscopy." American Society of Clinical Pathologists, Chicago.

Davis, B. J. (1964). Disc electrophoresis II: Method and application to human serum proteins. *Ann. N.Y. Acad. Sci.* 121, 404–427.

D'Amore, M. A., Gallagher, P. M., Korfhagen, T. R., and Ganschow, R. E. (1988). The complete sequence and organization of the murine β-glucuronidase gene. *Biochemistry* 27, 7131–7140.

Fishman, W. H., Goldman, S.S., and DeLellis, R. (1967a). Dual localization of β-glucuronidase in endoplasmic reticulum and in lysosomes. *Nature* (*Lond.*) 213, 457–460.

Fishman, W. H., Kato, K., Anstiss, C. L., and Green, S. (1967b). Human serum beta-glucuronidase; Its measurement and some of its properties. *Clin. Chim. Acta* 15, 435–47.

Funkenstein, B., Leary, S. L., Stein, J. C., and Catterall, J. F. (1988). Genomic organization and sequence of the Gus-s^a allele of the murine β-glucuronidase gene. *Mol. Cell. Biol.* 8, 1160–1168.

Gallagher, P. M., D'Amore, M. A., Lund, S. D., and Ganschow, R. E. (1988). The complete nucleotide sequence of murine beta-glucuronidase mRNA and its deduced polypeptide. *Genomics* 2, 215–219.

Glaser, J. H., and Sly, W. S. (1973). β-Glucuronidase deficiency mucopolysaccharidosis: Methods for enzymatic diagnosis. *J. Lab. Clin. Med.* 82, 969–977.

Glaser, J. H., Roozen, K. J., Brot, F. E., and Sly, W. S. (1975). Multiple isoelectric and recognition forms of human β-glucuronidase activity. *Arch. Biochem. Biophys.* 166, 536–542.

Haskins, M. E., Desnick, R. J., DiFerrante, N., Jezyk, P. F., and Patterson, D. F. (1984). β-Glucuronidase deficiency in a dog: A model of human mucopolysaccharidosis VII. *Pediatr. Res.* 18, 980–984.

Hayashi, M., Shirahama, T., and Cohen, A. S. (1968). Combined cytochemical and electron microscopic demonstration of β-glucuronidase activity in rat liver with the use of a simultaneous coupling azo dye technique. *J. Cell Biol.* 36, 289–297.

Himeno, M., Ohhara, H., Arakawa, Y., and Kato, K. (1975). β-Glucuronidase of rat preputial gland: Crystallization, properties, carbohydrate composition, and subunits. *J. Biochem.* 77, 427–438.

Jefferson, R. A., Burgess, S. M., and Hirsh, D. (1986). β-Glucuronidase from *Escherichia coli* as a gene-fusion marker. *Proc. Natl. Acad. Sci. USA* 83, 8447–8451.

Kornfeld, S., and Mellman, I. (1989). The biogenesis of lysosomes. Annu. Rev. Cell Biol. 5, 483–525.

Kyle, J. W., Birkenmeier, E. H., Gwynn, B., Vogler, C., Hoppe, P. C., Hoffmann, J. W., and Sly, W. S. (1990). Correction of murine mucopolysaccharidosis VII by a human β-glucuronidase transgene. *Proc. Natl. Acad. Sci. USA* 87, 3914–3918.

Lagunoff, D., Nicol, D. M., and Pitzl, P. (1973). Uptake of β-glucuronidase by deficient human fibroblasts. *Lab. Invest.* 29, 449–453.

Lowry, O. H., Rosebrough, N. J., Farr, A. L., and Randall, R. J. (1951). Protein measurement with Folin phenol reagent. *J. Biol. Chem.* 193, 265–275.

Lojda, Z. (1970). Indogenic methods for glycosides II. An improved method for β-D-galactosidase and its application to localization studies of the enzymes in the intestine and other tissues. *Histochemie* 23, 266–288.

Medda, S., and Swank, R. T. (1985). Egasyn, a protein which determines the subcellular distribution of β-glucuronidase, has esterase activity. *J. Biol. Chem.* 260, 15802–15808.

Miller, R. D., Hoffmann, J. W., Powell, P. P., Kyle, J. W., Shipley, J. M., Bachinsky, D. R., and Sly, W. S. (1990). Cloning and characterization of the human β-glucuronidase gene. *Genomics* 7, 280–283.

Neufeld, E. F., and Muenzer, J. (1989). The mucopolysaccharidoses. In "The Metabolic Basis of Inherited Disease" (C. R. Scriver, A. L. Beaudet, W. S. Sly, and D. Valle eds.), Vol. 2, pp. 1565–1587. McGraw-Hill, New York.

Nishimura, Y., Rosenfeld, M. G., Kreibich, G., Gubler, U., Sabatini, D. D., Adesnik, M., and Andy, R. (1986). Nucleotide sequence of rat preputial gland β-glucuronidase cDNA and in vitro insertion of its encoded polypeptide into microsomal membranes. *Proc. Natl. Acad. Sci. USA* 83, 7292–7296.

Oshima, A., Kyle, J. W., Miller, R. D., Hoffmann, J. W., Powell, P. P., Grubb, J. H., Sly, W. S., Tropak, M., Guise, K. S., and Gravel, R. A. (1987). Cloning, sequencing, and expression of cDNA for human β-glucuronidase. *Proc. Natl. Acad. Sci. USA* 84, 685–689.

Paigen, K. (1989). Mammalian beta-glucuronidase: Genetics, molecular biology, and cell biology. *Prog. Nucleic Acid Res. Mol. Biol.* 37, 155–205.

Powell, P. P., Kyle, J. W., Miller, R. D., Pantano, J., Grubb, J. H., and Sly, W. S. (1988). Rat liver β-glucuronidase: cDNA cloning, sequence comparisons and expression of a chimeric protein in COS cells. *Biochem. J.* 250, 547–555.

Sanes, J. R., Rubenstein, J. L. R., and Nicolas, J.-F. (1986). Use of a recombinant retrovirus to study post-implantation cell lineage in mouse embryos. *EMBO J.* 5, 3133–3142.

Sly, W. S., Quinton, B. A., McAlister, W. H., and Rimoin, D. L. (1973). Beta glucuronidase deficiency: Report of clinical, radiologic, and biochemical features of a new mucopolysaccharidosis. *J. Pediatr.* 82, 249–257.

Smith, B. F., Hoffman, R. K., Giger, U., and Wolfe, J. H. (1990). Genes transfered by retroviral vectors into normal and mutant myoblasts in primary cultures are expressed in myotubules. *Mol. Cell. Biol.* 10, 3268–3271.

Swank, R. T., and Paigen, K. (1973). Biochemical and genetic evidence for a macromolecular β-glucuronidase complex in microsomal membranes. *J. Mol. Biol.* 77: 371–389.

von Figura, K., and Hasilik, A. (1987). Lysosomal enzymes and their receptors. *Annu. Rev. Biochem.* 55, 167–193.

Vogler, C., Birkenmeier, E. H., Sly, W. S., Levy, B., Pegors, C., Kyle, J. W., and Beamer, W. G. (1990). A murine model of mucopolysaccharidosis type VII: Gross and microscopic findings in beta-glucuronidase-deficient mice. *Am. J. Pathol.* 136, 207–217.

Wolfe, J. H., Shuchman, E. H., Stramm, L. E., Concaugh, E. A., Haskins, M. E., Aguire, G. D., Patterson, D. F., Desnick, R. J., and Gilboa, E. (1990). Restoration of normal lysosomal function in mucopolysaccharidosis type VII cells by retroviral vector-mediated gene transfer. *Proc. Natl. Acad. Sci. USA* 87, 2877–2881.

APPENDIX

A Suppliers

Chemical

Biosynth AG
P.O. Box 125
CH-9422 Staad Switzerland
Phone: (071) 43 01 90
Fax: (071) 42 58 59
Telex: 882 929
For bulk (>1 gm) orders of X-Glucuro Chx and X-Glucuro Na.

Biosynth International, Inc.
P.O. Box 541
Skokie, Illinois 60076 USA
Phone: (708) 674-5160
Fax: (708) 674-8885
For bulk (>1 gm) orders of X-Glucuro Chx and X-Glucuro Na.

Boehringer Mannheim Corporation
Biochemical Products
9115 Hague Road
P.O. Box 50414
Indianapolis, Indiana 46250-0414 USA
Orders: (800) 262-1640
Technical Service: (800) 428-5433

Calbiochem Corporation
P.O. Box 12087
San Diego, California 92112-4180 USA
Phone: (800) 854-3417
(619) 450-5692
Fax: (619) 453-3552
Telex: 697934

Clontech Laboratories, Inc.
4030 Fabian Way
Palo Alto, California 94303 USA
Phone: (415) 424-8222
Fax: (415) 424-1352
Telex: 330060
Suppliers of the Gus reagents and fusion plasmids.

Fluka Chemical Corp.
980 South Second Street
Ronkonkoma, New York 11779-7238 USA
Phone: (516) 467-0980
(800) 358-5287
Fax: (516) 467-0663
(800) 441-8841
Telex: 96-7807

Hoefer Scientific Instruments
654 Minnesota Street
P.O. Box 77387
San Francisco, California 94107
USA
Phone: (415) 282-2307
(800) 234-4750
Fax: (415) 821-1081
Telex 470778

Richard Jefferson
CAMBIA
Center for the Application of
Molecular Biology to
International Agriculture
CAMBIA Organizational Office
Lawickse Allee 22
Wageningen 6707 AG
The Netherlands
Phone/Fax: +31-8370-26342

Molecular Probes, Inc.
P.O. Box 22010
4849 Pitchford Ave
Eugene, Oregon 97402 USA
Phone: (503) 344-3007
Fax: (503) 344:6504
Telex 858721 (Molecular)
Suppliers of GUS substrates.

Research Organics Inc.
4353 East 49th Street
Cleveland, Ohio 44125 USA
Phone: (800) 321-0570
(216) 883-8025 (Ohio)
Manufacturer of β-glucuronidase
substrates including Red-Gluc.

Sigma Chemical Company
P.O. Box 14508
St. Louis, Missouri 63178-9916
USA
Phone: (314) 771-5750
800-325-3010
Fax: (800)-325-5052
(314)-771-5757
Telex: 434475

United States Biochemical
Corporation
P.O. Box 22400
Cleveland, Ohio 44122 USA
Phone: (800) 321-9322
(216) 765-5000
Telex: 980718

Equipment

Beckman Instruments, Inc.
Spinco Division
1050 Page Mill Road
Palo Alto, California 94304 USA
Phone: (415) 857-1150
Telex: 678413

Bio-Rad, Inc.
3300 Regatta Boulevard
Richmond, CA 94804
Phone: (800) 227-5589 (outside
California)
(800) 277-3259 (inside
California)

Dynatech Laboratories, Inc.
14340 Sullyfield Circle
Chantilly, Virginia 22021 USA
Phone: (703) 631-7800
Continental USA: (800) 336-4543
Manufacturers of a microtiter plate reading filter fluorometer.

Flow Laboratories/ICN Biomedicals, Inc.
ICN Plaza
3300 Hyland Avenue
Costa Mesa, California 92626 USA
Telephone: (714)-545-0113
Fax: (714)-540-2080
Telex: 685580
Manufacturers of a microtiter plate reading filter fluorometer.

Hamilton Company
P.O. Box 10030
Reno, Nevada 89510
Phone: (702) 786-7077
TWX: 910-395-6031

Hitachi Instruments, Inc.
A subsidiary of Hitachi America, Ltd.
44 Old Ridgebury Road
Danbury, CT 06810
Manufacturers of a wide variety of fluorescence instrumentation.

Hoefer Scientific Instruments
654 Minnesota Street
P.O. Box 77387
San Francisco, California 94107 USA
Phone: (415) 282-2307
(800) 234-4750
Fax: (415) 821-1081
Telex 470778
Manufacturer of the TKO 100 filter fluorometer designed for GUS and DNA assays.

Knotes Biotechnology
1022 Spruce Street
Vineland, New Jersey 08360-2899 USA
Phone: (800) 223-7150
(609) 692-8500
Fax: (609) 692-3242

Kontron Instruments, Inc./ Research Instruments International
7915 Silverton Avenue, Suite 310
San Diego, California 92126
Phone: (619) 689-1100
Fax: (619) 689-1174
Manufacturer of fluorescence instrumentation.

Miles Scientific
Division Miles Laboratories, Inc.
2000 North Aurora Road
Naperville, Illinois 60566 USA
Phone: (800) 348-7465

Perkin-Elmer Corp.
761 Main Ave
Norwalk, Connecticut 06859
USA
Phone: (203) 762-1000
(800) 762-4000
Fax: (203) 762-6000
Manufacturers of a wide range of fluorescence instrumentation including a microtiter plate reading fluorometer.

Shimadzu Scientific Instruments, Inc.
7102 Riverwood Drive
Columbia, Maryland 21046 USA
Phone: (301) 381-1227
Fax: (301) 381-1222
Manufacturers of a wide variety of fluorescence instrumentation.

APPENDIX

B GUS Gene Constructs

Detailed In This Book

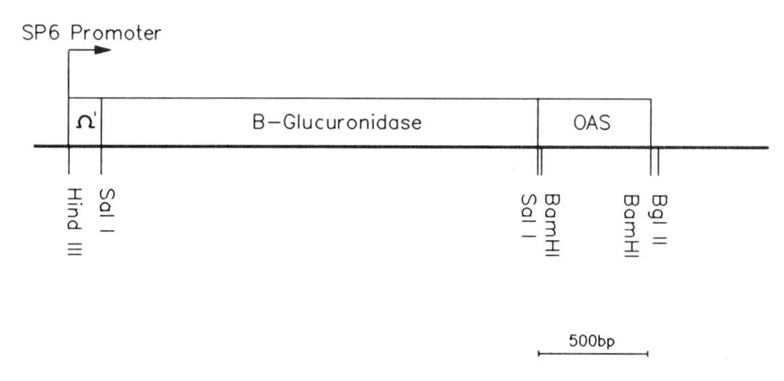

Osbourn and Wilson, Chapter 10, Figure 1 Map of SP6 transcription plasmid pJII140 (Sleat *et al.,* 1987). pJII140 was derived from pSP64 (Promega Corp.) via pSP64TMV (Sleat *et al.,* 1986), which contained a *Bam*HI fragment bearing the TMV origin-of-assembly sequence (OAS; genome coordinates 5118-5550). The GUS gene contains no common restriction sites (Jefferson, 1987). *Bgl*II-linearized pJII140 was transcribed as described in the text prior to incubation with TMV coat protein.

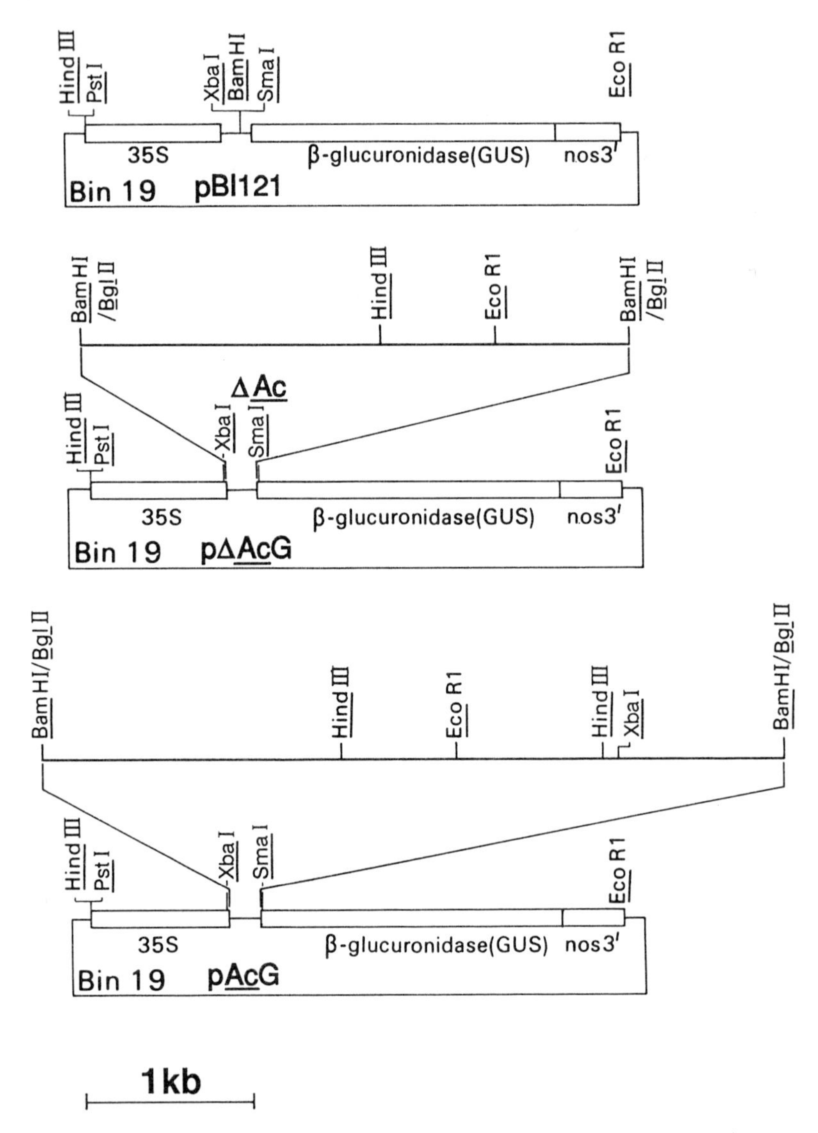

Finnegan, Chapter 11, Figure 2 Restriction maps for the binary vectors pBI121, pΔ*Ac* G, and p*Ac* G. Abbreviations: 35S, cauliflower mosaic virus 35S promoter; GUS, coding region of *E. coli* β-glucuroidase gene; nos3′,3′ termination region from nopaline synthase; Bin 19, binary vector (Bevan, 1984).

		GUS Specific Activity (nmoles/min/mg)	
		CHO	Tobacco
GUS	GpppG 17 b AUG —— GUS —— UGA 67 b —GGGUACCGAGCUCGAAUU-3'	1.4	0.01
GUS-$(A)_{25}$ DraI	GpppG 17 b AUG —— GUS —— UGA 67 b —$AU(A)_{25}UUU$-3'	20.0	0.34
GUS-$(A)_{50}$ DraI	GpppG 17 b AUG —— GUS —— UGA 67 b —$AU(A)_{25}GUU(A)_{25}UUU$-3'	30.0	0.71

Gallie *et al.*, Chapter 13, Figure 1 GUS constructs used for electroporation of CHO and tobacco cells. Approximately 1 μg of each construct was used for electroporation. See text for details.

Farrell and Beachy, Chapter 9 GUS constructs for protein targeting studies. See pGUSN358 → S in Clontech list below.

Commercially Available

Many of the following plasmids are used in this book, and all are available from Clontech laboratories. Alternatively, plasmids can be obtained by writing to Dr. R. A. Jefferson (see Appendix A) or to the authors of the appropriate work.

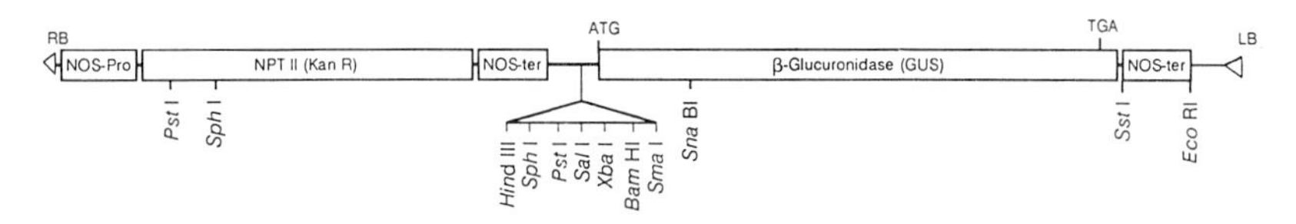

pBI101 (Jefferson *et al.*, 1987) Designed for testing promoter activity, pBI101 confers kanamycin resistance. pBI101 is a derivative of pBIN19 and is unstable unless grown in the presence of kanamycin.

pBI101.2 and pBI101.3 Identical to pBI101, except the reading frames are moved one and two nucleotides, respectively, relative to the polylinker.

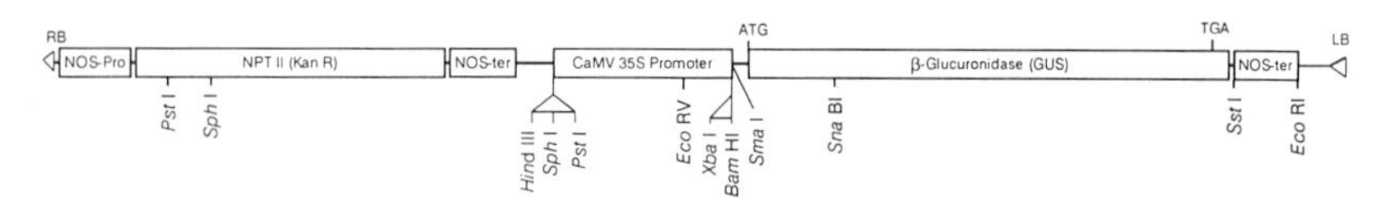

pBI121 (Jefferson *et al.*, 1987) A derivative of pBI101 containing the 35S promoter of the cauliflower mosaic virus.

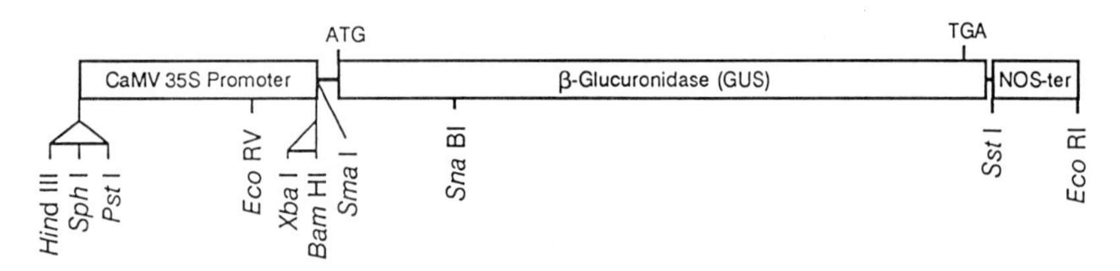

pBI221 The CaMV 35S promoter-GUS-NOS-ter portion of pBI121 was cloned into pUC19 to produce pBI221.

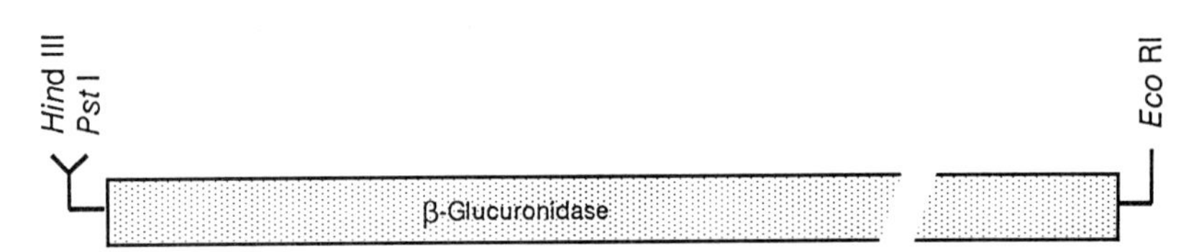

pRAJ255 (Jefferson *et al.*, 1986) A 1.87 Kb insert containing the GUS gene was cloned into pEMBL9 to produce pRAJ255.

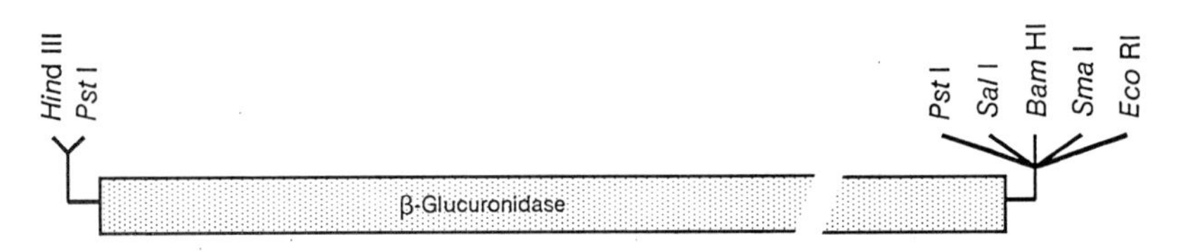

pRAJ260 Similar to pRAJ255 except for a modified Eco R1 site.

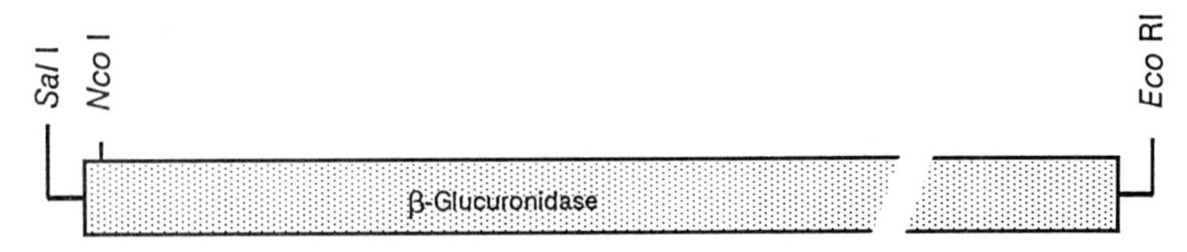

pRAJ275 This derivative of pRAJ255 contains a consensus translational initiator in place of deleted 5' GUS sequences.

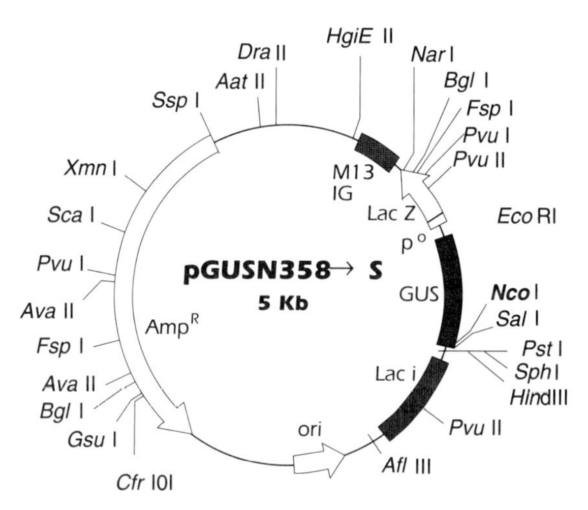

pGUSN358→S (Farrell and Beachy, Chapter 9, this book) pGUSN358→S contains a GUS gene modified by site directed mutatgenesis to eliminate a glycosylation site. This allows processing by the endoplasmic reticulum without the usual inactivation of GUS activity.

References

Jefferson, R. A., Burgess, S. M., and Hirsh, D. (1986). β-glucuronidase from *Escherichia coli* as a gene fusion marker. *Proc. Natl. Acad. Sci. USA* **86:**8447–8451.

Jefferson, R. A., Kavanagh, T. A., and Bevan, M. W. (1987). GUS fusions: β-glucuronidase as a sensitive and versatile gene fusion marker in higher plants. *EMBO J.* **6:**3901–3907.

Index

Page entries in italic indicate tables or figures.